I0392462

# The Least Action Nuclear Process Theory of Cold Fusion

## A Theory of Heat

Daniel S. Szumski, Independent Scholar

The Unusual F Street Winery Press
Davis, California

First edition : ICCF-20

# Author's Preface

I began my independent scholarship of reversible thermodynamic processes over 35 years ago. I wanted to explore what I believed was a cleaver way to model a living cell...modeling only its internal energy storage rather than the molecular quantities. I knew that the cell's energy storage must be as far-from-equilibrium as were its bio-molecular quantities. I also knew that I had to understand free energy, and soon thereafter I understood that what I wanted to do required a non-equilibrium Theory of Heat; one which brought together in a single equation the heat radiation theory of Max Planck, and the heat of molecular motion in a gas, championed by James Clerk Maxwell and Ludwig Boltzmann. I was reasonably well positioned for this task. My physics education is self taught, but more important, entirely classical. I have read all of the English accessible benchmark works by Boltzmann, Maxwell, Helmholtz, Planck, Wien, Gibbs, Ehrenfest, Einstein, Schrödinger, Szilárd, Shannon and Weaver, and Prigogine to mention only those that were most important to my Theory of Heat. The resulting cell model had three degrees of freedom, and no molecular quantities. It describes only the highly organized structure of radiation domain energy in the cell's covalent bonds. The model is particularly interesting in the way that it treats the living cell as a reversible thermodynamic process.

I soon found out that no one was interested in listening to a civil engineer talk about reversible processes, and even less so when the discussion crosses over into a Theory of the Living State.

The person who gave me a 'read' was William Hoover at Lawrence Livermore Labs. His words went something like this: "Your theory of heat has merit, but this is 19$^{th}$ century science. You need to find a more contemporary application, some place where there is a lot of ongoing theoretical discourse". I remembered those words as I was reading a news article about cold fusion. Here was a theoretical area that cried out for what I had developed; a Theory of Heat. I was surprised by the speed with which my theory explained how Mev levels of reversible process energy could be accumulated in a Fleischmann and Pons electrode. But my understanding went further, describing why the loading time was so great, and where the energy might be stored. The theory found energy accumulation occurring in very small, radio frequency quanta.

So, being the naïve engineer, I began writing down the broadest range of nuclear fusion reactions that could be derived from isotopes contained in George Miley's (U. of Illinois) nickel electrode. These reactions were all thermodynamically possible, but the energy requirements were extraordinary and unlikely, but that didn't stop the naïve engineer. It was particularly satisfying to draw the decay sequences for each of my transmutation products, and then calculate the mass/energy change for each stable end product. My comparisons to Miley's transmutation table found that a large fraction of the stable end products were not observed in Miley's final electrode.

However, my study of classical thermodynamics had taught me that all reversible processes had outcomes that were precisely and unambiguously determined by the Principle of Least Action alone. And as I went through my decay calculations, I was stunned that the stable isotope products having the smallest mass/energy change were all in Miley's final electrode, and there were no false positives. That is, I didn't predict any isotope that was not in Miley's table.

At that time, I didn't realize that I had a more ominous problem before me. Everyone knew that reversible thermodynamic processes don't actually occur in Nature. I had to convince hard nosed scientists, some of the best in the world, that some variant of perpetual motion was possible, and that it was the common denominator in understanding Nature's two most guarded secrets: cold fusion, and life.

My independent scholarship has led me to the opinion that the single greatest scientific discovery of the 21$^{st}$ Century will be an understanding of what I call the *imperfect reversible thermodynamic process*, a process that is flawed, but at such a small energy departure from the perpetual motion criteria, that the reversible process can maintain its integrity in an otherwise irreversible thermodynamic world. When it is fully developed, this theory will provide profound insights into the two scientific questions that continue to elude theoretical explanation; our cold fusion process, and matter's living state. Both are very far-from-equilibrium events that exist at the farthest limit of the Second Law of Thermodynamics. My current understanding of where the imperfection might be located is to be found in Chapters 1 and 8. It appears to be at the lower end of the statistical distribution around the mean kinetic energy value, $k_BT$.

For those who want to read about the LANP model's explanation of specific laboratory observations, I am providing this quick reference glossary:

| | | |
|---|---|---|
| 1. | Conformance to the Laws of Thermodynamics | Chapter's 5 and 8 |
| 2. | The Reason for Long Loading Times | pg. 62 |
| 3. | Source of Energy Needed for Ignition | pp. 22-23, 40-42, 52, 62 |
| 4. | Circumventing the Coulomb Barrier | pg. 53 |
| 5. | Absence of Gamma Radiation | pg. 43-44 |
| 6. | Thermonuclear Temperatures | Chapter 5 |
| 7. | Thermonuclear Energy Storage | pg. 52 |
| 8. | Why Thermonuclear Conditions are Hidden from our Observation | pp. 11, 53, 59 |
| 9. | The Ordering Mechanism for Stable Nuclear Transmutation Products | pp. 27-29, 69-71 |
| 10. | High Performance Electrode Design | Chapter 9 |

One final note: I want to challenge those who call this venture into reversible thermodynamics, cold fusion and living systems: fringe-science research. If it is, I am in good company. Two other respected cold fusion scientists; John O.M. Bockris and Jean Paul Biberian have published learned papers in both the cold fusion literature and the biological transmutation literature. This proximity of interests is not accidental. I believe that we will find that both processes involve extraordinarily high internal temperatures that are capable of sponsoring nuclear transmutations, and both will be found to be *imperfect reversible thermodynamic processes*.

Dan Szumski, Independent Scholar
Davis, California, danszumski@gmail.com
www.LeastActionNuclearProcess.com
September 15, 2016

# Table of Contents

## Chapter 1

# What is a Reversible Thermodynamic Process?

## A. Introduction

A reversible thermodynamic process accomplishes work without the expenditure of energy. It is perpetual motion.

We know that such a process cannot exist in the natural world. Max Planck is most frequently quoted in this regard: "As Helmholtz has pointed out, all these reversible processes have the common property that they may be completely represented by the principle of least action, which gives a definite answer to all questions concerning any such measurable process, and to this extent (the) theory of reversible processes may be regarded as completely established. Reversible processes have, however, the disadvantage that singly or collectively they are only ideal: in actual nature there is no such thing as a reversible process"(). Helmholtz was Planck's mentor.

This might seem like an incredible indictment of this books purpose. However, in time we will see that it is not. It is merely the starting point. Let's begin.

### B. The Direction Of Time's Arrow

In order to begin to answer our question: 'What is a reversible thermodynamic process?' we need to understand time. Time is the progression of an entropic event. We experience it as the event sequence that always proceeds in one direction, and generally at a constant speed. It is an integral part of the theory that we hope to develop. However, our understanding of time needs to be clarified in two respects.

The first time measure that we want to explore is directional. Our positive time direction will be that in which the Second Law of Thermodynamics (the great law of entropy increase) prevails, and order is always and everywhere going over into disorder. This is the time direction wherein a weight falls according to Newtonian principles, celestial objects move and evolve, chemical reactions proceed to new equilibrium conditions, and (this is the important one in what is to follow) where dielectric loss continually changes three-dimensional electrical charge into the heat of three-dimensional, atomic and molecular motion.

But, more importantly, we will concern ourselves with the opposite time direction, a negative direction that when combined with our positive direction, modifies it, and slows the entropic time progression to where it perceptively slows, or even seems to stop. This is the time direction that dominates in living systems and gives them what Erwin Schrodinger described as life's neg-entropic character, that is, a series of events that proceeds to progressively more ordered states.

Secondly, we will want to explore the progression of events, using as a measure of time, its inverse...the frequency domain in which discrete events take place. We will use this as our measure of perception. It defines the time scale in which perceived events take place, and by extension, the sequence in which those events occur. This sequence can be either exact, as in the case of DNA transcription, or containing a random component, in a statistical-mechanical or thermodynamic sense. For example, randomness might describe in probabilistic terms, the statistical distribution of deuteron velocities near the surface of a metal hydride's crystal lattice. Here too, we will be adding something entirely new, and this is important, ...a method for defining, from among all of the possible statistical mechanical events that can occur, precisely which one will occur next.

But first, let's look at how time is relative depending upon the event that is occurring. Let's choose as an example, a chemical event, and in particular, the redox event: $2[H^+] + O_2 + 2\{e^-\} \xrightarrow{v_f} H_2O_2$, where the forward velocity of the reaction is $v_f = k_f[H^+]^2[O_2]\{e^-\}^2$, and $k_f$ has the units per time, and $v_f$, the forward velocity of the reaction, is measured in $moles^5/time$.

However, this is only half of the overall reaction:

$$2[H^+] + [O_2] + 2\{e^-\} + e^{H_f'/k_bT} \underset{v_b}{\overset{v_f}{\longleftrightarrow}} [H_2O_2] + e^{H_f''/k_bT}$$
$$2[H^+] + [O_2] + 2\{e^-\} \underset{v_b}{\overset{v_f}{\longleftrightarrow}} [H_2O_2] + e^{\Delta H_f/k_bT}$$

where: $v_b = k_b[H_2O_2][\exp(\Delta H_f')]$

and $K_{eq}^o = v_f/v_b = \frac{k_b[H_2O_2][\exp(\Delta H_f')]}{k_f[H^+]^2[O_2]\{e^-\}^2}$, the true equilibrium constant for the reaction. If we then set: $v_b = v_f$, $K_{eq}^o = 1.0$,

and the apparent equilibrium constant, $K_{eq}$:

$$K_{eq} = \frac{k_f[e^{-\Delta H_f/k_bT}]}{k_b} = \frac{[H_2O_2]}{[H^+]^2[O_2]\{e^-\}^2} \text{, } \log K_{eq} = 23.1, E_H^o = 0.68 \text{ volts}$$

The positive value of $\log K_{eq}$ indicates a reaction that is heavily favored in the forward time direction. However, while the forward reaction may dominate, there is a small, but still significant reaction in what we would see as the negative time direction. In other words, we see a chemical event that results in $H_2O_2$ production in our positive time direction. But in looking more closely at the process, this is the net event, where a portion of its overall structure occurs in negative time, or the direction of entropy decrease. This reaction produces equilibrium states, and is purely entropic. We will call this the *normal entropic process*.

### Forward, Backward and Net Velocity

$2[H^+] + [O_2] + 2\{e^-\} \underset{v_b}{\overset{v_f}{\longleftrightarrow}} [H_2O_2]$

Becomes:

$2[H^+] + [O_2] + 2\{e^-\} \xrightarrow{v_f - v_b} [H_2O_2]$

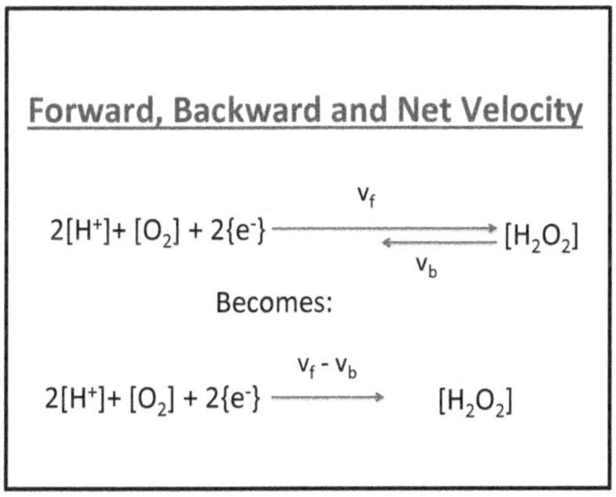

In an idealized world, it should be possible to alter the reaction to make the backward direction more dominant. This can be accomplished by altering either, or both velocities to make the difference between them smaller. This slows the overall reaction, and effectively decreases its entropy production. Schrodinger would say that we have made the process move in the negative entropy direction, or more simply, that it has become *negentropic*, a state in which the net reaction velocity is positive, but smaller than that of the *normal entropic process*. As the neg-entropic state develops, it moves farther-from-equilibrium, toward more and more improbable thermodynamic states, and increasing instability. From the time that the process left the normal entropic state it was a *negentropic process*.

Now look what happens when we take this negentropic process to its limit; allowing the forward and backward reaction rates to become identical. This special case, places the process at the very limit of what the Second Law allows. Entropy production ceases, and the process is precisely balanced among all of its possible outcomes. There exists an equal probability of evolving in the forward and backward time dimensions, and all mass/energy and energy conversion outcomes are possible. Time appears to stand still. This is as far-from-equilibrium that the process being considered can go in the negentropic direction. We call this limiting case of the negentropic process, the *reversible process*.

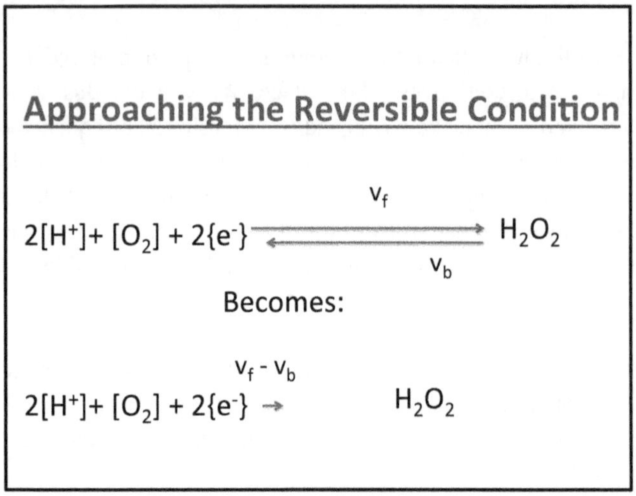


However, the importance of the reversible process to cold fusion science lies beyond these circumstances. This is because all reversible processes, regardless of their nature, be it mechanical, electro dynamical, chemical, or electromagnetic, have one additional constraint that bestows on them their very special place in science. It is the Principle of Least Action. And what makes it so indispensible is the precision with which it specifies from among all of the possible 'next steps' that the reversible process could evolve to, the one 'next step' that results in the *least action*. And if the process remains in its reversible state after this step, there is again only one 'next step' that the process can evolve to, and it too, is determined by the Principle of Least Action. In this way, we see how any process that remains in a reversible thermodynamic state, must trace out a very specific temporal evolution that we might refer to as the *Least Action Process*. It does not matter that this reversible process occurs as an isolated process, or as a small part of a much larger statistical mechanical process. As long as the process evolves within the framework of reversible thermodynamics, every 'next step' is precisely and unambiguously determined by the Least Action Principle, and at least in theory, its complete temporal evolution is precisely deterministic.

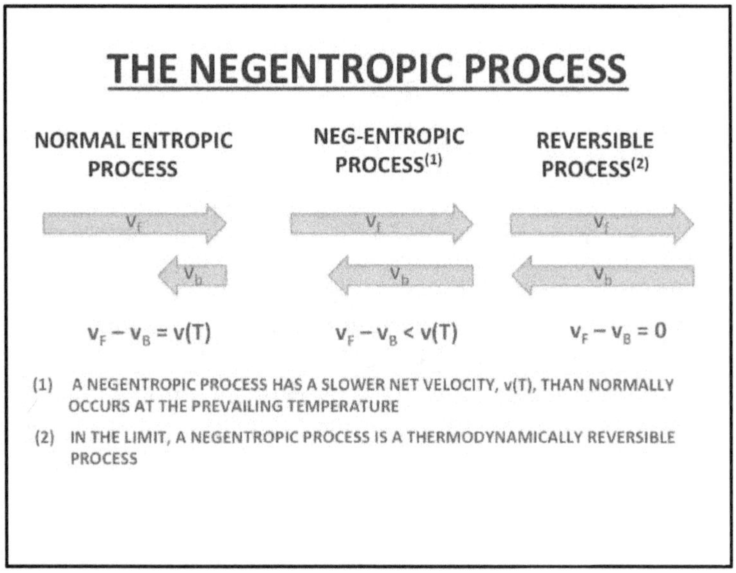


## C. The Reversible State Is Not A Reversible Process

Reversible states are common in Nature, and while they are frequently mistaken for a reversible process, they are fundamentally different. Consider a glass tube containing molecular oxygen. In its ground state the oxygen molecules in the tube are randomly oriented and have energies that are statistically distributed around a temperature dependent mean energy, $k_bT$, in accordance with the Maxwell-Boltzmann distribution. If we now place this tube in a magnetic field, the paramagnetic oxygen molecules, which become tiny magnets, tend to orient in the direction of the magnetic field. This is called a reversible state because when the magnetic field is removed, the oxygen molecules return to the random orientations of their original ground state.

Do you see how we have introduced magnetic energy to drive the orientation of the oxygen molecules to a very far-from-equilibrium state, and then allowed the system to relax into its ground state by removing the magnetic field? But this is not a reversible process. It has accomplished no net work, and indeed, energy has been expended to impose and then remove the magnetic field.

**D. The Reversible Process in Nature** [Discussion until the end of this chapter contains references to concepts explained in later chapters of this book]

Now consider the reversible process. Let's use as an example, a covalent bond. Biological systems create these bond by localizing free energy in the form of a single photon that is resident between the bond's covalent electrons; being emitted by one, absorbed by the other, emitted by that one, and absorbed by its covalent partner, and so on. Work has been introduced into this system to: 1) bring the covalent electrons into proximity, and 2) initiate the process. However, the force that brought about this bond is no longer necessary, and the covalent process continues its function, without energy renewal, indefinitely. Ilya Prigogine would call this a *dissipative structure*, a far-from-equilibrium displacement to a state from which there is no return to the previous state. All reversible thermodynamic processes begin in this manner.

Once the displacement to that second state is complete, the covalent bond can be moved to another location, or placed in proximity to other forces, and it remains unchanged. It exhibits no energy expenditure, or energy change of any kind. And yet, it continues indefinitely. No entropy increase occurs.

This places the covalent process at the very limit of the Second Law. The work that it accomplishes is the chemical bond between those two electrons (and their atoms). And that work continues indefinitely without energy input or losses of any kind. We see this in the covalent bonds that persist in preserved biological specimens, ancient wooden artifacts and even in DNA extracted from preserved dinosaur tissue. Furthermore, any of these samples can be burned to release the energy contained in their covalent bonds. The covalent bond is an example of a *perfect reversible process.*

A photon traveling in the vacuum of space is a second example of a reversible thermodynamic process. It requires an energy displacement in an electron from a higher to a lower energy state, which causes the photon to be emitted in the first place, but once its energy is dissipated into the photon state, the initial forcing function is no longer necessary for the reversible photon process to continue. Instead, the photon takes on a life of its own. The work that it accomplishes is the transport of energy in space and time. It exhibits no expenditure of that energy, nor does it require energy renewal for its continuance. It has no losses, and its entropy remains unchanged. This too, is a *perfect reversible process.*

In both of these examples, free energy is relocated into the reversible process state where it resides as radiation domain energy, a photon; and it continues to perform work indefinitely. This happens because at every next step in the process the energy change is identically zero. The photon's initial and final states are always identical.

It is important to understand that in both of these reversible processes there is no kinetic energy, and the process is exactly deterministic. These are distinguishing characteristics of the reversible thermodynamic process. In the first case, kinetic energy is a mass domain property. It brings with it statistical uncertainty and statistical mechanical behavior. There is no such thing as strict determinism when material particles are involved, particularly when those particles exist at the scale of atoms or molecules. In such a system there is always Heisenberg Uncertainty, and the analytical methods of statistical thermodynamics are the only quantitative method. As these minute particles are aggregated into biological proteins or crystalline structure their kinetic energy is quieted, and their behavior becomes less statistical and more deterministic.

Reversible thermodynamic processes are, on the other hand, strictly deterministic. Heisenberg Uncertainty cannot exist, and for this reason, the reversible process cannot involve material particles and their kinetic behavior. In fact, all reversible thermodynamic processes that we will be considering exist solely in the radiation energy domain, or the radiation domain. Recall the photon, or the photon resonance within a covalent bond.

When mass particles are present in the reversible process space, they have no kinetic energy relative to the photon energy, and the radiation domain energy acts to quiet the mass particles in opposition to their tendency for thermal motion, and its statistical characteristics. Consider the covalent bond. Its reversible process is exclusively in the photon that is continually absorbed and emitted by the covalent electron particles. But the electrons do not move relative to one another. Their motion relative to the reversible photon process is identically zero. In addition, the photon energy and its position are precise and exactly deterministic. There are two distinct electrons, and two distinct atoms involved in this covalent bond. The probability of their participation is identically one. And while the molecule where this union occurs may move around, there is no relative movement of the mass particles in the reversible process space.

### E. The Imperfect Reversible Process

In the cold fusion model that we will be studying there will be an antecedent step that will make the reversible process imperfect. The initial energy will be derived from the kinetic energy of deuterons...mass domain particles. The theory proposes that the dissipative structure is the transfer of kinetic energy from the mass domain into the radiation domain, and that this transfer must take place within the reversible process space. This is because reversible processes allow this kind of energy transfer from one type to any other type; irreversible thermodynamics does not.

If this does indeed occur, there must be an exception to the *no kinetic energy rule* of reversible processes. That is an upper limit on the amount of kinetic energy that can exist within the reversible process space without impeding its forward progression. And if we want to be more specific about the numerical criteria that we are seeking, it appears to be an upper limit on the reversible process loading rate itself, an upper limit on the kinetic energy that can be loaded in a single reversible process step.

So what does the reversible process loading step look like? We know that the process is strictly deterministic which allows us to do something that is not possible in processes governed by statistical thermodynamics. In particular, we can zero in on the activity of a specific deuteron and quantize its thermodynamic behavior. Our deuteron's kinetic energy, $1/2\,mv_d^2$, goes to zero, thereby removing an infinitesimal amount of energy-of-motion from the heat reservoir adjacent to the uptake site, and reducing the thermodynamic temperature there. This temperature differential is being transferred into a corresponding change in the radiation domain temperature. However, if the process is to continue, the temperature/energy deficit outside the deuteron absorbance site must be renewed. Do you see how the reversible process has to operate in an imperfect way to load energy? The absorbed energy needs to be continually replaced. This energy amount must be minute(mi-noot) to slide beneath the reversible process radar, and is likely below $k_bT$. But how much below?

The kinetic energy of the deuteron ensemble is statistical in nature, having a mean value estimated by $k_bT$. However, from among all of those deuteron kinetic energies, the reversible process selects the *lowest* energy deuteron, in accordance with the Principle of Least Action. This kinetic energy absorption is the thermodynamic cost of the imperfect reversible thermodynamic process. It has essentially harvested that cost from the heat reservoir immediately outside the absorption site, where it is easily restored from the heat reservoir within the cold fusion device. Does this make it clear that there is a way for an imperfect reversible thermodynamic process to operate in an otherwise irreversible world?

### D. The Reversible Process is Strictly Deterministic

All reversible thermodynamic processes are governed by the Principle of Least Action, wherein every next step in, for instance, the cold fusion process is precisely and unambiguously determined because it results in the smallest energy change. This has two implications. First, the process is strictly sequential. Another next step can only occur only when the previous one is finished. This places finite limits on how fast transmutations can occur within a specific electrode. Additionally, it probably limits weaponization because energetic cascades like those in fission reactors and nuclear weapons cannot occur.

Do you see how this also makes the process strictly deterministic? Once the operative least action measure is isolated, as for instance the smallest energy change in the nuclear transmutation reaction space, it becomes possible to array all of the possible candidate next steps and choose from among them the one that will occur. The products resulting from that step are then added to the inventory of candidate next steps, and the next prediction takes place. This is how one might go about optimizing a high performance cold fusion electrode for excess heat or isotope production. Chapter 9 discusses this and other electrode design considerations.

There is one final operating characteristic of the reversible thermodynamic process that specifies the conditions under which it can take place in nature. Remember how this Chapter began with a statement by Max Planck regarding the impossibility of reversible thermodynamic processes in nature. And yet, I cite two above: the photon in the vacuum of space, and the covalent bond. In both of these processes, the next steps are repetitive and identical. They are also known precisely. We can say with certainty which next step is the least action step, and that its energy change will be identically zero. These are the defining characteristics of a *perfect reversible thermodynamic process*.

## E. The Reversible Process Leaves No Trace In Nature

It is not possible to measure or even observe a reversible process. It has no outward manifestation...no losses...nothing is absorbed or emitted. The reversible process is however detectible in its interactions with the irreversible world. The photon imparts its energy when an electron absorbs it. The covalent bond releases its energy when the material is burned.

This characteristic is evident in the cold fusion process in the way that thermonuclear temperatures and energies are hidden from our view. We know that they are there because of the observed nuclear transmutations. The First Law requires that thermonuclear energies be present for these transmutations to occur. Chapter 7 describes the physics that allows these extraordinary energies in a room temperature device.

## F. Reference

(1) M. Planck, Verhandlunger der Deutschen Physikalischen Gesellschaft, 2, 237, (1900), or in English translation: Planck's Original Papers in Quantum Physics, Volume 1 of Classic Papers in Physics, H. Kangro ed., Wiley, New York (1972).

# Chapter 2

# A Theory of Far-From-Equilibrium Heat Storage

## A. Introduction

Planck's blackbody emittance equation (1) is the universally accepted model for heat radiation's equilibrium, spectral distribution. It has been found superior to any other contemporary form (2,3,4,5,6). However, this acceptance is only justified for equilibrium, and leaves two important issues unresolved. First, Planck's solution provides no insight into non-equilibrium or far-from-equilibrium states, or the mechanisms of redistribution between equilibrium states (7). Only Ehrenfest (8) has explored redistribution mechanisms, and Forte (9) describes a non-equilibrium Wien Displacement Law.

Secondly, Planck's energy quanta violated the continuity requirements of Maxwell's equations. Einstein first enunciated this discrepancy, and Planck spent the next 2 decades, unsuccessfully trying to resolve it. Meanwhile, the experiments of Stark (10), and Einstein's light particle theory (11) demonstrated the dual nature of light, galvanizing the discontinuity's place in physics. More recent studies of non-quantum blackbody theory (12,13,14,15,16,17) have not reconciled this conflict.

This paper explores one possible avenue to a non-equilibrium blackbody equation. The goal here, is understanding free energy partitioning between the domains of heat radiation, and molecular motion. One of the model's solutions is then interpreted as an avenue to understanding far-from-equilibrium energy storage in living cells and cold fusion devices.

## B. Theory

Planck (18) viewed the separation of all physical phenomena into reversible and irreversible processes as the most elemental, and most important, because all irreversible processes share a common similarity that makes them unlike any reversible process. This distinguishing characteristic is the transformation of heat energy to motion, which can in no way be referred back to the process from which it came. This research considers Maxwell's electromagnetic wave traveling undiminished in time, its information content preserved, until it encounters a material particle.

Light absorption is considered a two-step process. The first is an adiabatic reversible step, wherein one-dimensional light energy is absorbed in a quantum amount, $h\nu$, by an electron, and is wholly contained within it. The absorbed quantum is still 1-dimensional (1D), remains within the domain of reversible thermodynamics, and does not emit Joule heat. There is no recourse to the Second Law during this first step.

The absorption process' second step is a dimensional restructuring that the 1-D electrical quantum undergoes in evolving into its 3-D equivalent, electrical charge density. This occurs in accordance with the Equip-partition Theorem, along the axes of the electron's three spatial coordinates. The resulting displacement of the generalized coordinates translates to 3-D motion, the evolution of Joule heat, and irreversibility. The magnetic vector has no 3-D equivalent, and can only transform to 1-D paramagnetic spin. Accordingly, photon de-coupling distorts time's fabric, giving rise to the characteristic spectral emittance.

This second absorption step is represented by integrating the Dirac delta over the 3-D transformation's time interval

(1) $$\int_{t_1}^{t_2} \frac{1}{t}\,dt = \ln\left({t_2}/{t_1}\right) = \ln\left({\upsilon_1}/{\upsilon_2}\right) \;[\textit{frequency domain}]$$

with variance distribution:

(2) $$\sigma^2_{\upsilon_1/\upsilon_2} = \left[\ln\left({\upsilon_1}/{\upsilon_2}\right)\right]^2 = f_1\left({\upsilon_1}/{\upsilon_2}\right)$$ 1st Integral

The probability distribution's re-structuring, $\nu_m$ from a 1-D quanta, $h\nu$, to three dimensions requires taking the next two integrals of the variance function relative to the Wien frequency, $\nu_m$. Thus:

(3) $$f_n\left({\upsilon_1}/{\upsilon_m}\right) = \iint \left[\ln\left({\upsilon_1}/{\upsilon_m}\right)\right]^2 d\left({\upsilon_1}/{\upsilon_m}\right)$$

The results are summarized as:

(4) $$f_2\left({\upsilon_1}/{\upsilon_m}\right) = \left({\upsilon_1}/{\upsilon_m}\right)^1\left[\left(\ln\left(\frac{\upsilon_1}{\upsilon_m}\right)\right)^2 - 2\,\ln\left(\frac{\upsilon_1}{\upsilon_m}\right) + 2\right]$$ 2nd Integral

(5) $$f_3\left({\upsilon_1}/{\upsilon_m}\right) = \left({\upsilon_1}/{\upsilon_m}\right)^2\left[\frac{1}{2}\left(\ln\left(\frac{\upsilon_1}{\upsilon_m}\right)\right)^2 - \frac{3}{2}\,\ln\left(\frac{\upsilon_1}{\upsilon_m}\right) + \frac{7}{4}\right]$$ 3rd Integral

The third integral represents all of the possible interconnections between any arbitrary frequency and the Wien frequency. $\nu_m$is the most probable frequency at the prevailing temperature. $\nu_1$ is the damped frequency continuum of the blackbody spectra.

Assuming the radiation absorption function to be exponentially distributed,

(6) $$a_\upsilon = e^{f_3(\upsilon_1/\upsilon_2)}$$

and substituting this absorption and the Raleigh-Jeans emittance into Kirchhoff's Law:

(7) $$K(\upsilon_1) = \frac{e_\upsilon}{a_\upsilon} = k_b T_m \cdot \frac{\upsilon_1^2}{c^2} \cdot \frac{1}{e^{f_3(\upsilon_1/\upsilon_m)}}$$

Blackbody Spectra = Emittance (Rayleigh Law) / Absorptance (This study)

This spectral distribution: 1) is derived entirely from classical theory; 2) contains the discontinuity indirectly, $\nu_m(h)$; 3) incorporates the consequences of electro-magnetic theory (Rayleigh-Jeans Law); and 4) suggests a mechanism for exchange of energy between frequencies. Sears [19] gives Wien's Displacement Law as:

(8) $\nu_m = 5.89\times10^{10} \cdot T_R(^oK)$ (Wien frequency)

The Figure displays features of the blackbody radiation spectra described in this way. The figure also displays calculations using Planck's Equation:

(9)

$$K'(\nu_1) = h\nu_1 \frac{\nu_1^2}{c^2} \cdot \frac{1}{e^{\frac{h\nu_1}{k_b T}} - 1}$$

The agreement is close, but not exact, differing by less than 2% at $\nu_m$, and in the 5%-8% range to the left, where Eq. (7) better represents Raleigh-Jeans. The curves terminate at the point where the calculations yield partial quanta. The number of quanta is obtained by dividing the spectral emittance by the conversion factor $\nu^2/c^2$ and then dividing by $h\nu$.

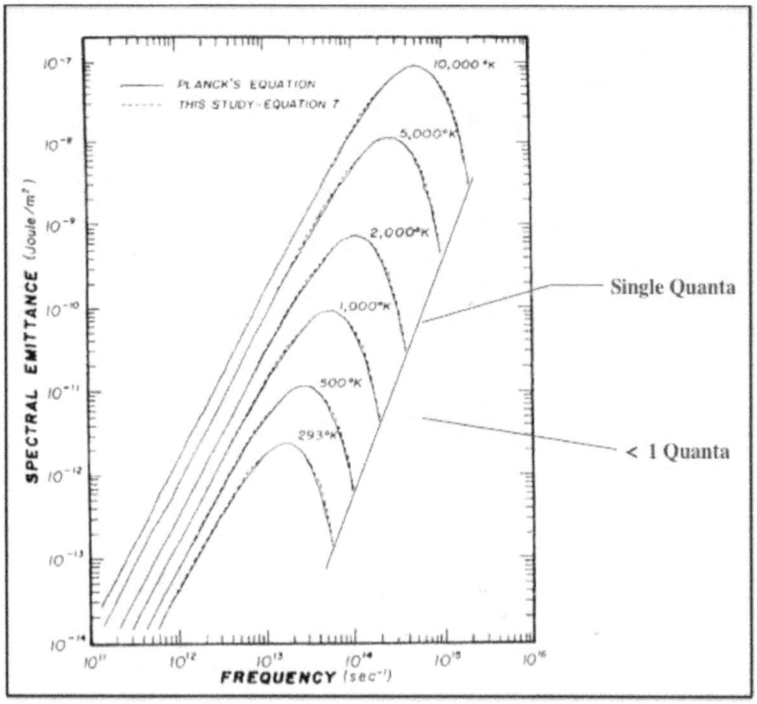


## C. Discussion

Both the two-slit experiment and the photoelectric effect are consistent with this theory. The wave properties of light are unaltered. The theory eliminates discontinuity as a property of light, placing it instead in the electron's absorption of light, or more precisely, in the electrons absorption of light only in quantum amounts. This reassignment isn't contrary to, nor does it change, existing theory.

Eq. (7) offers two significant advances over Planck's, which are instructive in furthering our understanding of heat processes. The first is Eq. (7)'s explicit statement for energy transference between frequencies. This was identified at the outset as the distinguishing characteristic of the required non-equilibrium blackbody form. Eq. (5) suggests that the common channel for energy re-distribution is the Wien frequency, since each spectral frequency is explicitly related to it. Planck's equation can also be shown to contain the same ratio (21).

Second, Eq. (7) contains two distinct thermodynamic scales, representing the entire range of non-equilibrium heat conditions. The concept of two temperature scales is not new (22,23,24,25,26,27). The first of these scales is the classical thermodynamic temperature, of the Rayleigh-Jeans Law,$T_m$. It is common to both equations, and expresses the temperature of thermal motion alone.

The second temperature, that contained in the Wien Displacement Law, is identical to the first where the system is in equilibrium. However, it is fundamentally different from $T_m$ in ways that could give profound meaning to Eq. (7). This is the radiation temperature, $T_R$. That it can be expressed in the same units as the classical thermodynamic temperature is seen in the equilibrium case. However, changes in $T_R$, independent of the thermodynamic temperature, shift the spectral distribution in plausible non-equilibrium ways that may provide insight into both non-equilibrium and far-from-equilibrium heat processes.

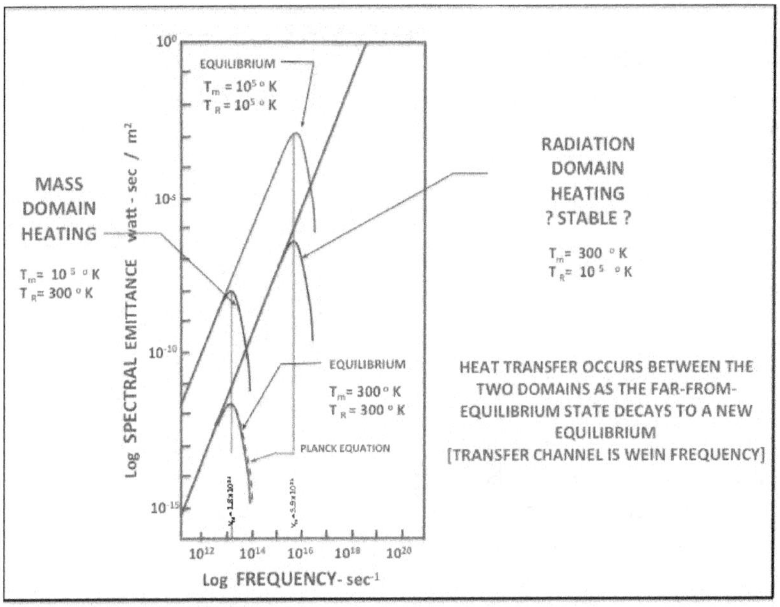


In the next figure, $T_R$, and consequently the Wien frequency, remains constant while $T_m$ increases from 300 $^\circ K$ to $10^5$ $^oK$ (Case A). This represents a sudden frictional input of heat to a material body that is initially at thermal equilibrium. Similarly, $T_R$ can be increased without a corresponding increase in the thermodynamic temperature (Case B). The radiation density within the blackbody is increased without a corresponding increase in the Rayleigh-Jeans emittance. The new region delineated by this spectral distribution consists primarily of higher energy radiation, but the process from which it arises appears to an observer to be adiabatic, and might therefore, be viewed as completely reversible. From this theory's standpoint, the energy content within this new region (Case B) consists entirely of radiation transfers that are undergoing the first stage of radiation absorption, alone. That is, radiation is fully absorbed in its one- dimensional form and immediately re-emitted. There is no de-coupling of light's electro-magnetic structure, and therefore no entropy increase. This is the initial condition when high-energy radiation strikes a body initially at equilibrium.

Taking this result further, one might ask: Are there states in nature that exploit the energy/entropy relationship suggested by these calculations? There might be. Living systems are constructed of high energy covalent bonds that both, represent very far-from-equilibrium conditions, and store larger amounts of electro-magnetic energy than would normally exist at the $T_m$. It is possible that Case B shows how far-from-equilibrium energy storage might be masked from ambient thermodynamic conditions in matters living state. Each covalent electron pair shares the wave function, $\psi^2$, alternately absorbing and re-emitting light energy, but only in a manner consistent with this theory's first absorption step. This portion of the heat radiation spectra is localized (masked) between electron pairs, and does not contribute to either the measurable heat spectra or to dielectric losses. Thus, the thermodynamic temperature of the cell ($T_m$) is unaffected, and a stable far-from-equilibrium condition with lower localized entropy, is possible. The degree of entropy decrease is defined by the separation between $T_m$and $T_R$. The permanence of that change appears to depend on irreversible storage of neg-entropy outside mechanistic pathways back to equilibrium (28). As the energy storage requirements begin to exceed the capacity of the covalent bond system, other mechanisms for storing this biological energy in a stable far-from-equilibrium state might be found in covalent bond, and probably even excited nuclear states, as Mossbauer resonance. Eq. (5) suggests enormous capacity for far-from-equilibrium entropy absorption and the information storage this implies.

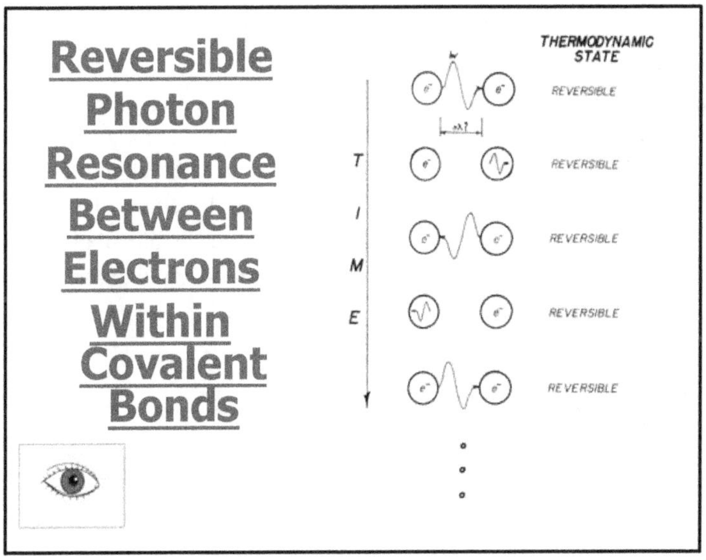

A second example of where this type of heat theory may prove important to scientific understanding is found in what has been called cold fusion. Energy storage in a palladium or nickel electrode might occur in the radiation domain (Case B in Figure 3), first as excited electronic states and a corresponding increase in the redox state, and later in excited nuclear (Mossbauer) states. Is it possible that such a system could exhibit radiation temperatures approaching $10^7\ ^oK$ while the ambient temperature of the electrolysis apparatus hovers around $330\ ^oK$ ? Stranger paradigms have occurred in the history of science.

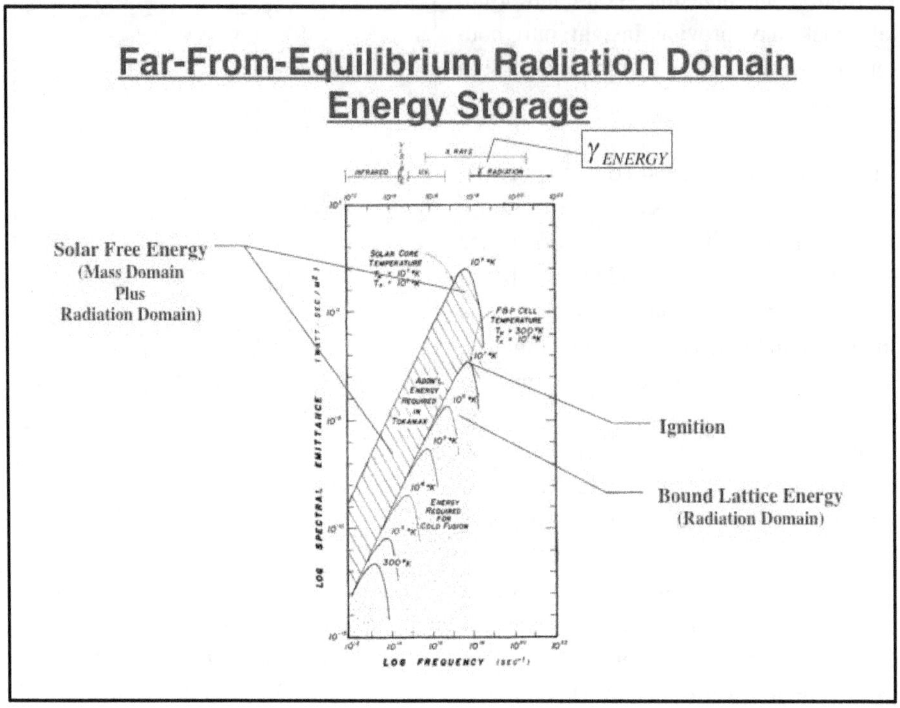


A third example of far-from-equilibrium spectral energy might be found in sono-luminescence, wherein mechanical energy is converted to electro-magnetic energy. In this case, mechanical energy increases $T_m$ instantaneously without a corresponding increase in $T_R$ (Case A in Figure 3). Lacking any mechanism to maintain this far-from-equilibrium condition, the system spontaneously moves toward equilibrium by channeling the stored mechanical energy through the Wien frequency channel, and thence, into the radiation domain. If the energy flux is high enough, visible light is observed.

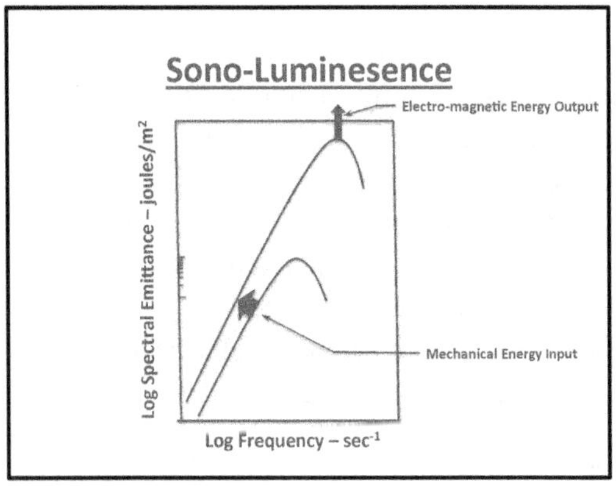


The 1-D to 3-D transform function given by Equation 5 could possibly be a mathematical statement of the Second Law at the boundary between electrodynamics and mechanics. In its temporal form (19) the equation represents the relative dominance of the forward (entropic) and backward (negentropic) reaction directions. The entropic direction is that in which waste heat of motion is produced, while the negentropic direction is characterized by increased order where the radiation domain predominates.

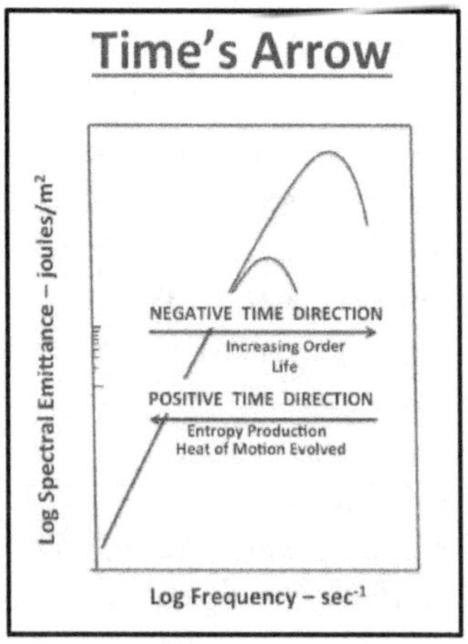

## C. References


(1) M. Planck, Verhandlunger der Deutschen Physikalischen Gesellschaft, 2, 237, (1900), or in English translation: Planck's Original Papers in Quantum Physics, Volume 1 of Classic Papers in Physics, H. Kangro ed., Wiley, New York (1972).

(2) H. Rubens and F. Kurlbaum, Ann. Physik, 4, 649 (1901).

(3) F. Paschen, Ann. Physik,4, 277 (1901).

(4) E. Warburg, Ann. Physik, 48, 410 (1915).

(5) W. Nernst and T. Wulf, Ber. deut. phys. Ges., 21, 294(1919).

(6) W.W. Coblenz, Dict. Appl. Phys., Vol. IV, "Radiation".

(7) T.S. Kuhn, Blackbody Theory and the Quantum Discontinuity 1894-1912, Oxford University Press, New York (1978).

(8) Kuhn found references to non-equilibrium transitions only in Ehrenfest's notes.

(9) J. Fort, J.A. Gonzalez, J. E. Liebot, Physical Letters A, 236,193-200 (1997).

(10) J. Stark, Phys.ZS.,8(1907), 913-919, received 2 December 1907.

(11) A. Einstein, Ann. d. Phys.,17(190), 132-148, 1905.

(12) H. Dingle, Phil Mag, XXXVII,246, 47 (1946).

(13) T.H. Boyer, Phys. Rev. 182, 1374 (1969).

(14) T.H. Boyer, Phys. Rev. 186, 1304 (1969).

(15) T.H. Boyer, Phys. Rev. D, 11, 790 (1975).

(16) T.H. Boyer, Phys. Rev. D, 29, 1089 (1984).

(17) D.C. Cole, Phys. Rev. A, 45, 8471 (1992).

(18) M. Planck, Eight Lectures in Theoretical Physics-1909, translated by A.P. Wills, Columbia U Press, NY (1915).

(19) F.W. Sears and M.W. Zemansky, University Physics, 3rd ed., Addison-Wiley, Reading, MA (1964).

(20) See T. Preston, Theory of Heat, Macmillan and Co, London (1929) for a description of their methods and results.

(21) If we rearrange Equation (8) and substitute for T in Planck's Equation (9), the exponential term becomes $e^{2.8276\left(v_1/v_m\right)}$.

(22) B.C. Eu, L.S. Garcia-Colin, Phys. Rev. E, 54, 2501 (1996).

(23) D, Jou, J. Casas-Vazquez, Phys. Rev. A, 45, 8371 (1992).

(24) K. Henjes, Phys. Rev. A, 48, 3194 (1993).

(25) W.G. Hoover, B.L. Holian, and H.A. Posch, Phys. Rev. A, 48, 3191 (1993).

(26) D. Jou, J. Casas-Vazquez, Phys. Rev. A, 48, 3201 (1993).

(27) J. Fort, D. Jou, and J.E. Leobot, Physica A, 269, 439(1999).

(28) G. Nicolis, I. Prigogine, Self-Organization in Non-Equilibrium Systems, John Wiley and Sons, NY, 1977.

Chapter 3

# The Least Action Nuclear Process Model of Cold Fusion

## A. Introduction

During the last two decades it has become evident that low energy nuclear reactions are occurring in Fleischmann-Pons (F-P) electrolytic cells (1). These reactions are unprecedented in nuclear physics, and are at least for now, hidden from understanding because a suitable theoretical framework has not been forthcoming. (2)

It is theorized that excess heat (3,4) and ${}^4He$ (5) are generated, and that the heat evolved is consistent with the mass difference (5,6) in the reaction:

(1) $(2){}^2_1H^+ \Rightarrow {}^4_2He + 23.9\,MeV$ (ignition requirement= 0.01MeV)

It is also becoming apparent that the reactions taking place are a near surface phenomenon (7) that is spatially clustered (7), occurs in bursts (7), and has a cyclic character within those bursts (8). Even more controversial than the contention that Reaction (1) occurs, is a growing body of data showing other nuclear transmutations in F-P cells (9,10,11,12,13), and still more alarming, in living cells (14).

**'New Science' is Required**

The degree to which new physics underlies these experimental observations is not known. But, among theoreticians it is considered more likely that the present conundrum will be resolved by extensions of known physical principles, perhaps in ways that we cannot immediately imagine.

This research endeavors to provide insight into three theoretical issues. First, recognizing that the fusion reaction's energy has its origin within the experimental apparatus, we explore a mechanism for accumulating the energy required by Reaction (1) or other fusion/fission reactions. Second, there has to be a way of storing that energy within the apparatus, and in some way, disguising it until the moment of ignition. And the third is the elusive coherence principle that focuses the accumulated energy on specific nuclear transformations, and not others. The goal here is to show how a different view of heat processes, one that includes both irreversible and reversible thermodynamics, might inspire a comprehensive cold fusion theory.

## B. Theory of Heat

Heat exists in two domains that continually exchange energy as any arbitrary thermal system tends toward new quasi-equilibrium states. These are the domain of molecular motion, and the domain of heat radiation. The first might be referred to as the mass domain. Its description was first formalized by Maxwell (15), and then by Boltzmann (16). Their theory represents the molecular velocity distribution of an ideal gas as a function of the system's temperature and the gas molecules' mass. It is an equilibrium theory stating the functional dependence of temperature and thermal motion. It was Helmholtz who had first shown that molecular motion is equivalent to heat; an observation that is central to what follows. Max Planck, in his 1909 lectures at Columbia University (17), elevates this insight to an equal footing with Maxwell's treatment of light as electromagnetic waves.

Heat energy also exists in the radiation domain. The theoretical framework describing equilibrium conditions there bears the revered names of Rayleigh, Wien, and Planck. Planck's equation (18) describes the equilibrium temperature dependence of blackbody spectral emittance.

Reversible thermodynamic processes are believed to be rare in nature. These are processes that produce a net zero free energy change, and are described by the thermodynamic treatment of Helmholtz, but not that of Gibbs. In all cases, reversible processes can be completely described by the Principle of Least Action. A discussion of this principle and the thermodynamics of reversible processes are presented by Planck (17).

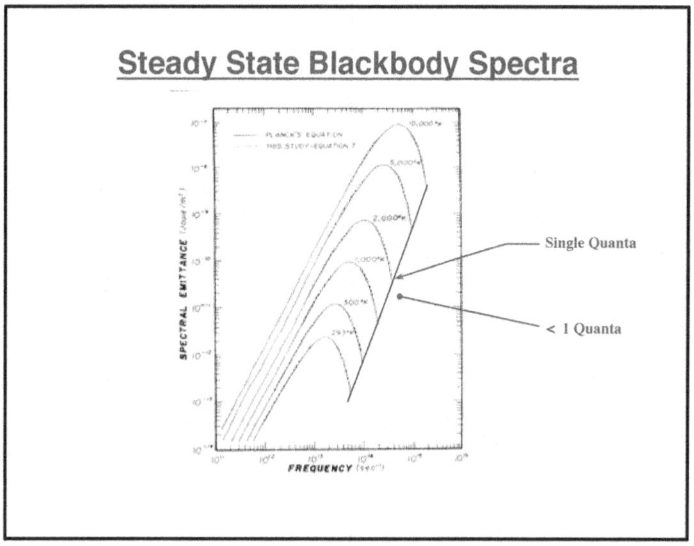


In a previous paper (19), I proposed a mathematical form for the blackbody spectral distribution that permits a glimpse into its non-equilibrium, and far-from-equilibrium characteristics. The theory treats light absorption as a two step process: the first (and this will be the important one to what follows) being wholly reversible, the second irreversible and entropic. In special cases only the first step occurs, and the energy absorption is adiabatic, that is, without loss of Joule heat. The functional form of the non-equilibrium blackbody spectra is given by:

(2) $$K(\upsilon_1) = \frac{e_\upsilon}{a_\upsilon} = k_b T_m \cdot \frac{\upsilon_1^2}{c^2} \cdot \frac{1}{e^{f_3(\upsilon_1/\upsilon_m)}}$$

Blackbody Spectra = Emittance (Rayleigh Law) / Absorptance (This study)

where

(3) $$f_3\left(\upsilon_1/\upsilon_m\right) = \left(\upsilon_1/\upsilon_m\right)^2\left[\frac{1}{2}\left(\ln\left(\frac{\upsilon_1}{\upsilon_m}\right)\right)^2 - \frac{3}{2}\ln\left(\frac{\upsilon_1}{\upsilon_m}\right) + \frac{7}{4}\right]$$

(4) $$\nu_m = 5.89\times10^{10} \cdot T_R \ ({}^oK) \quad \text{(Wien frequency)},$$

and $K(\nu_1)$ exists where the number of quanta is equal to or greater than 1.

Planck's equation for the equilibrium case is given by:

(5) $$K'(\nu_1) = h\nu_1 \frac{\nu_1^2}{c^2} \cdot \frac{1}{e^{\frac{h\nu_1}{k_b T}} - 1}$$

The non-equilibrium form of the equation is close to, but not exactly Planck's at equilibrium. The difference between the two may itself be of fundamental importance in understanding entropy increase in nature.

Secondly, the theory suggests that independent temperature scales might represent the mass and radiation domains. The figure illustrates the principle characteristics of the non-equilibrium, or more accurately the far-from-equilibrium, blackbody radiation spectra. Two equilibrium cases are shown: $300^oK$ and $100{,}000^oK$. Curve A, labeled Mass Domain Heating, refers to the transient initial condition where heating is initiated by increasing molecular motion, for example, by frictional input of heat. The Wien Frequency remains constant momentarily, and there is a logarithmic increase in the spectral energy at all frequencies.

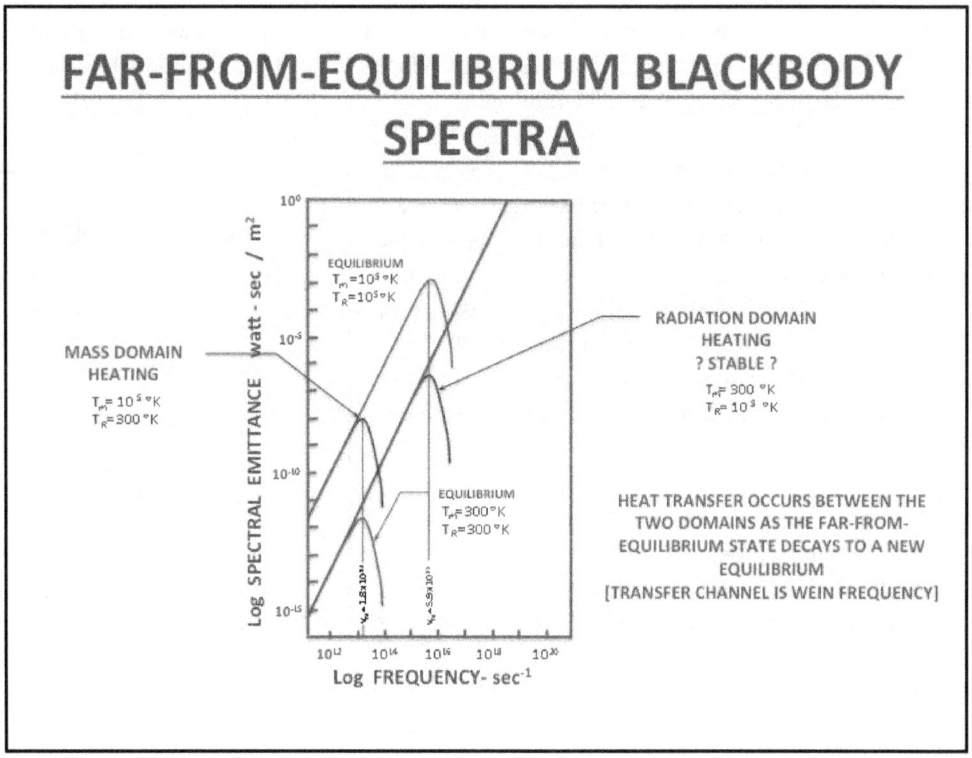


If an identical amount of heat is instantaneously added via the radiation domain alone (by introducing higher energy radiation), the thermodynamic temperature, $T_m$, is initially constant, and the Wien frequency increase, shifts the emittance spectra to higher frequencies as shown in curve B.

Both of these cases decay to an equilibrium spectrum similar to, but with higher total energy than, the initial equilibrium case. At the new equilibrium condition, the mass and radiation temperatures become identical, and it is not possible to determine from which of the two domains the original heating took place. However, as will be shown in what follows, there are circumstances under which the second of these spectra might be held in its far-from-equilibrium condition, and in this way store vast amounts of energy in a nickel or palladium cathode that is apparently at about 100 $^oC$.

This paper attempts to reconcile this non-equilibrium theory of heat with experimental observations from cold fusion experiments, and show how the energy stored at room temperature can be made accessible to fusion reactions.

## C. Consider A Thermodynamically Reversible Process

In thermodynamically reversible chemical processes all of the available energy, including that of thermal motion, is utilized, and none of the energy in the post-reaction space is lost to random thermal motion. I have found that the easiest way to understand the principles involved in what is to follow, is by considering the reaction occurring at the active site of a biological enzyme. The reactants and enzyme have fundamentally different thermodynamic properties. The former are relatively small molecular forms having both chemical potential and thermal motion. The enzyme, on the other hand, is a large molecule, that in its globular active form has very little or no thermal motion, and an undetermined chemical potential.

The enzyme mediated biological reaction brings reactant molecules into conformational position, generally by electro-static attraction, and in so doing, makes improbable reactions, probable. The secret lies in transformations to the energy states of both the reactants and the enzyme. In particular, cleavage at the active site eliminates thermal motion in the reactants; it quiets them. The First Law tells us that the 'lost' thermal energy must be conserved, and in its limit, the Second Law tells us under what conditions the reaction can proceed. In essence, the reactants' thermal motion has become part of the reactant-enzyme complex, elevating its overall free energy content.

If the thermodynamics of the enzyme/reactant complex are truly reversible, the total energy is passed on to the reaction products, and there is no energy residual that contributes to thermal motion in the reaction space. As long as these conditions are met, the reaction proceeds in accordance with the Principle of Least Action. Then conformational changes occur, and the product becomes subject to the slightly altered thermal state of its environment. In essence, the enzyme has harvested random heat motion from the environment, converted it to useful work, and in so doing increased the radiation domain's heat content. What at first appears to be a violation of the Second Law, is simply its limiting case, a zero net energy reaction that produces a more negentropic state.

It was Szilard's argument (20) concerning Maxwell's sorting demon (15) that correctly showed how the negentropy stored by the demon as molecular organization and intellect, sponsors his trick, in apparent violation of the Entropy Principle. In the case considered here, it is the massive information content in the globular enzyme form that allows the demon to operate. In a related theoretical context, Prigogine (21) would label the enzyme/reactant complex a dissipative structure: a far-from-equilibrium thermodynamic state, which once formed, allows no recourse to the previous state, and in so doing, lowers the local entropy.

## D. Consequences of Deuterium Absorption into a Metallic Lattice

Consider a deuterium ion, $({}_1^2H^+)$, in Fleischmann and Pon's original experiment (1). Its total energy is the sum of its chemical potential and its kinetic energy, $\varepsilon$, that associated with temperature dependent random motion. The kinetic energy is given by the product of the deuteron's mass, $m_d$, and its temperature dependent velocity squared:

(6) $$\varepsilon = \frac{m_d(v(T_m))^2}{2}$$

Where: $m_{d=} m_p + m_n = 3.34\times10^{-27}\ kgm$, and $v$ is the velocity of an individual deuteron, or in our simplified treatment, the average velocity of an ensemble of deuterons at F-P cell temperature, $T_m$. We will assume an average velocity of 0.2m/sec, a simplification that ignores for the moment the system's actual velocity distribution, but facilitates illustrative calculations.

The average kinetic energy of the deuterons in their F-P cell can be calculated as

(7) $$\bar{\varepsilon} = 6.68\times10^{-29}\ Joules = 6.68\times10^{-22}\ ergs.$$

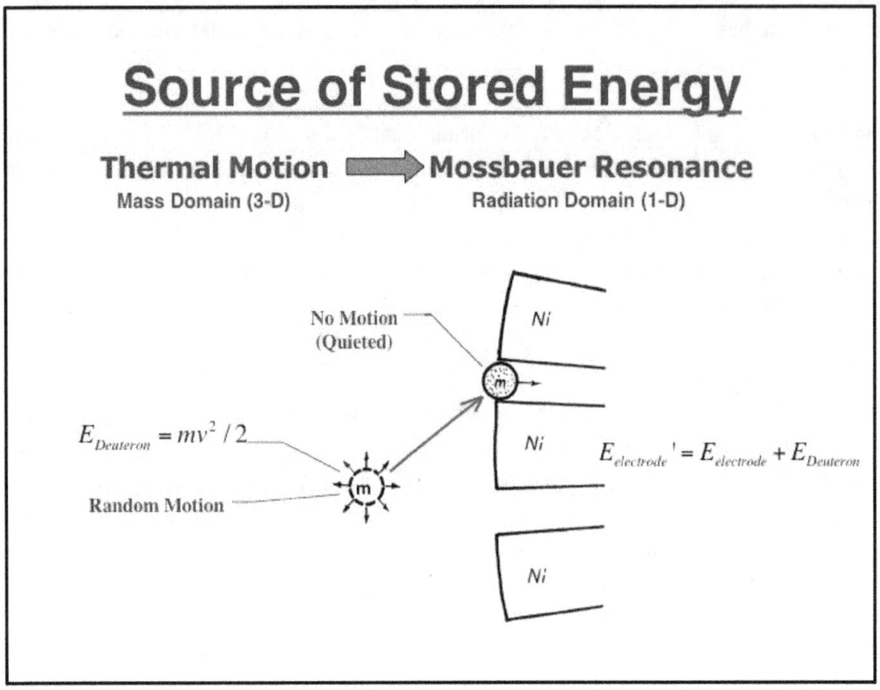


When a deuteron first encounters the nickel matrix, it is absorbed into it in a process that we will assume to be thermodynamically reversible, and similar to the enzyme process described above. The deuteron is 'quieted' to zero velocity, and zero kinetic energy. The First Law requires that the kinetic energy be conserved in the metal hydride lattice. And because the loading process is thermodynamically reversible, the energy storage is adiabatic, with no losses to Joule heat.

To place an order of magnitude estimate on this energy storage, we will use the 0.2cm diameter x 10cm electrode from Fleischmann and Pons 1989 experiments (1). The surface area of the cathode is $6.28\times10^{-4}$ meters. Assuming $\beta$-phase absorption approximating $\beta - PdD_{0.85}$, and having a lattice parameter of 0.405nm, the number of filled sites at the surface of the cathode, $\xi$, is approximated as:

(8) $$\xi = {6.28\times10^{-4}m^2}/{(0.405\times10^{-9}m)^2} = 3.83\times10^{15} \text{ surface sites}$$

The 85% load factor yields: $3.25\times10^{15}$ $[{}^2_1H^+]$ sites on a single atomic layer at the cathode surface. We assume that the total cathode is immersed in heavy water.

The energy storage capacity, E, of only the surface layer of atoms in this cathode is:

(9) $$E_{surface} = \bar{\varepsilon}\cdot\xi = 2.2\times10^{-6} ergs = 1.35MeV,$$

which is more than sufficient to ignite the fusion reaction:

(10) $$(2)\,{}^2_1H^+ \xrightarrow{fusion} {}^4_2He + \gamma_{23.9MeV} \quad \text{(ignition requirement = 0.01MeV)}$$

Absorption of deuterons into the second, third, and deeper atomic layers in the cathode, increases the total energy availability proportionally. For example, ignition is achieved in a single cathode layer if the average deuteron velocity is reduced to 2.0cm/sec, and in 100 layers if the velocity is further reduced to 0.2cm/sec. Energy accumulation increases at higher temperatures, and doubles again if ${}^2_1H_2$ molecules form within the interstitial space (22). Thus, deuterium's sequestered thermal motion appears to be more than sufficient for ignition.

## E. Possible Modes of Energy Storage within a Metallic Ni Lattice

The mechanism presented thus far has the advantage of providing qualitative insights into several theoretical issues. First, it provides a simple explanation of how the ignition energy is first acquired in the Ni cathode. All that might be required is a reversible thermodynamic process that harvests kinetic energy from the F-P cell environment. Secondly, it provides a basis for understanding the 'breathing' mechanism in the SRI experiments (8). Stored thermal energy and deuterium are expended and need to be replaced on a periodic basis. This could manifest as a harmonic superimposed on the excess heat output. Third, it offers an explanation of the apparent surface nature of the effect. This is where the energy accumulation occurs, and where it is most probably utilized and renewed. And, finally, it provides a plausible explanation of why loading rates increase at higher current density/temperature. The total energy storage per mole of ${}^{2}_{1}H^{+}$ is increased as the square of the average deuteron velocity.

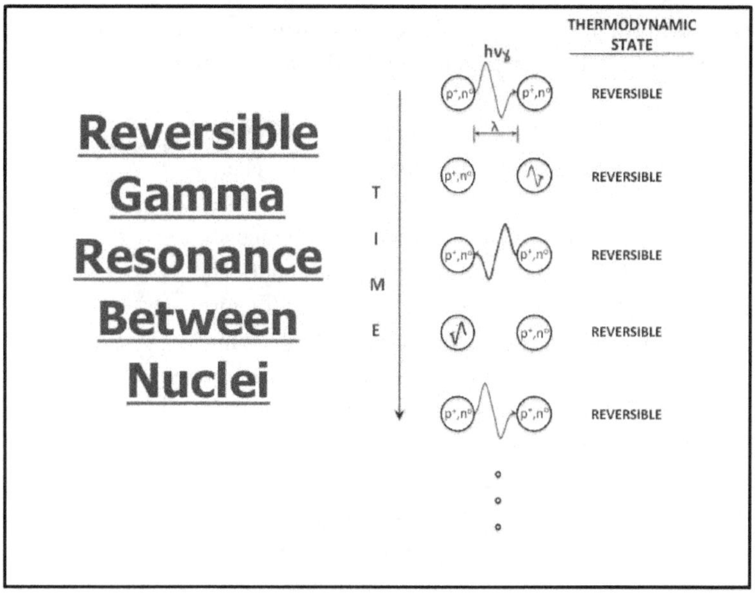


For this transfer to be a thermodynamically reversible one, the lattice energy has to increase sequentially, in discrete amounts, exactly equal to each sequestered deuteron's total kinetic energy. Then it must be held there in opposition to all entropic tendencies until ignition.

How is this energy stored during the loading phase of the experiment? We will begin by assuming that deuterium loading is a singular, multi-site, reversible process. The energy is not dissipated in incremental amounts, as it was in the enzyme example. Instead, the Ni lattice has massive numbers of active sites that must be filled before the reversible process can proceed further. During this loading, no energy is lost to thermal motion. Thus, the energy stored is either entirely in the radiation domain, or it moves from the mass and radiation domains of heat energy to another energy type where it can be held in a completely reversible state.

This constraint, that the energy storage be in a thermodynamically reversible state, allows us to further limit the possibilities. The mode of energy storage could be 1) electro-magnetic, in which case the energy of, for example, discrete metallic bonds might be increased by quantum amounts forming covalent bonds or excited electronic states; or 2) it could be magnetic energy storage in paramagnetic Pd's electron spin re-orientation, or 3) it could be energy stored as excited nuclear states. It is not stored as elastic stress, electric charge, or atomic vibration, all of which are entropic processes. In addition, the energy storage mechanism must make allowances for energy storage that spans a continuous range from the ambient temperature of the experimental apparatus, through thermonuclear temperatures.

It appears to me that the best explanation for the lower bound might be found in energy storage within discrete covalent bonds; each covalent electron pair alternately absorbing and emitting electro-magnetic energy that remains in a wholly reversible state, i.e. the first step of the two step, photon absorption process. This is a stable far-from-equilibrium state (upon which excited electron states might be superimposed). It manifests as an increase in the cathode's redox potential. As the total energy storage increases further excited nuclear states become active, ultimately bringing the reversibly stored cathode energy to gamma levels, where Mossbauer resonance, a reversible process, prevails, and energy storage occurs as resonant gamma exchange.

If we now look more closely at the consequences of energy storage in excited nuclear states, we find that this energy is stored entirely within the atomic structure of the lattice, and without any external manifestation. No heat energy is emitted. The thermodynamic temperature remains unaffected by the deuterium loading, and in this way, the process's energy storage is masked from observation. The observer witnesses a very typical electrolysis apparatus, and has no hint of the continually increasing radiation temperature within the lattice's atomic structure.

The spectra labeled B in the Reversible Heat Storage figure (pg 20) represents the distribution of energy levels corresponding to this storage of heat energy. These are filled sequentially at each Wien frequency. Then the Wien frequency increases one unit, and another layer is added to the spectral structure. Eventually, the Wien Frequency reaches gamma intensities, and the radiation temperature approximates that in the solar core, about $10^{7\,o}K$ as illustrated in the figure. The figure contrasts the temperature regime ($T_m$ and $T_R$) that this theory postulates, to that in the solar core. It suggests that the energy spectra required for ignition in the Tokamak includes both the indicated radiation domain energy and also the mass domain kinetic energy, and is about four orders of magnitude higher than that operative in the F&P cell. The total energy requirement is many orders of magnitude greater. In essence, the cold fusion process takes an energy shortcut around the enormous kinetic energy required for thermonuclear fusion. In this way, we see that the cold fusion process is actually quite hot.

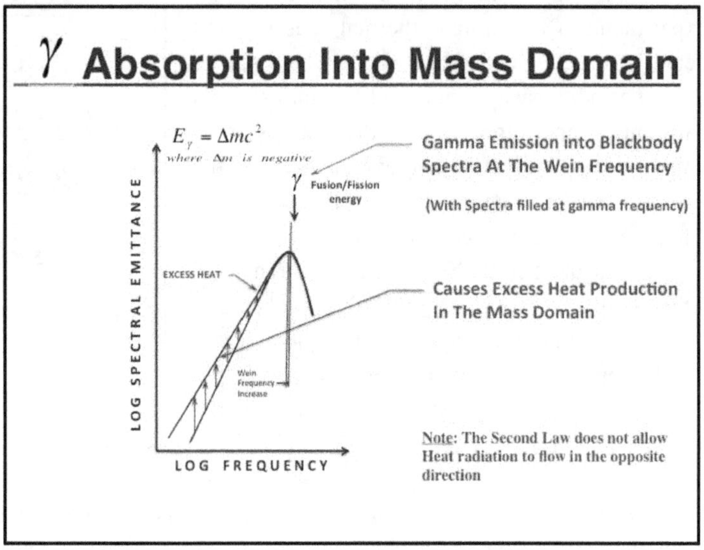


Let us, for the moment, assume that fusion and/or fission events occur in our electrode as $T_R$ approaches $10^{7\,o}K$. Where are the Gamma emissions?

We can answer this question by recalling that we are dealing here with an extension of blackbody theory wherein electromagnetic energy of all wavelengths is emitted and fully absorbed within the lattice. The mass quantities involved in this absorption and emission are electrons at the low energy end of the spectrum, and atomic nuclei as the energies increase through gamma intensities. This is an absorption/emission process wherein electro-magnetic energy is shared between identical mass quantities. In effect, gamma emission occurs as part of the normal blackbody dynamic, and within that context, participates in nuclear fusion or fission events. Gamma absorption and emission adds and subtracts energy quanta to/from the spectrum. Because this is a metal lattice, the emission/absorption occurs in accordance with Mossbauer kinetics, without recoil or heat loss, or more precisely in a completely thermodynamically reversible manner. And, as long as there is room in the spectra, the gamma energy released by fusion and fission events is fully absorbed elsewhere in the lattice by a nucleus having exactly the same ground/excited state as the emitting nucleus.

## Where are the Gamma Emissions?

- **Blackbody spectra - All wave lengths are absorbed and emitted**
  - Low Energy - Covalent Bonds
  - High Energy - Mossbauer Resonance
- **Mass Changes occur as Emission/Absorption in Blackbody Spectra**
- **Gamma Emission and Absorption are:**
  - Intrinsic to this System
  - Internal to It
  - In Effect, Masked by It

This is simple blackbody behavior, occurring now, at a very far-from-equilibrium state. One might rightfully ask: is it the mere fact that the gamma portion of the blackbody spectra is being

**Typical Nuclear Reactions**

$$^{58}_{28}Ni + (3)^{2}_{1}H^+ \xrightarrow{Fussion} {}^{64}_{31}Ga \xrightarrow{\beta^-} {}^{64}_{30}Zn$$

| | | |
|---|---|---|
| Mass | 63.977648 amu | 63.929142 amu |
| Mass Change | | -0.048506 amu |

$$^{58}_{28}Ni + {}^{64}_{28}Ni - (2)^{4}_{2}He \xrightarrow{fusion} {}^{114}_{52}Te \xrightarrow{\beta^+} {}^{114}_{51}Sb \xrightarrow{\beta^+} {}^{114}_{50}Sn$$

| | | |
|---|---|---|
| Mass | 113.858102 amu | 113.902779 amu |
| Mass Change | | +0.044676 amu |

$$^{114}_{50}Sn \xrightarrow{fission} (2)^{57}_{25}Mn \xrightarrow{\beta^-} {}^{57}_{26}Fe$$

| | |
|---|---|
| Mass | 113.870788 amu |
| Mass Change | +0.012685 amu |

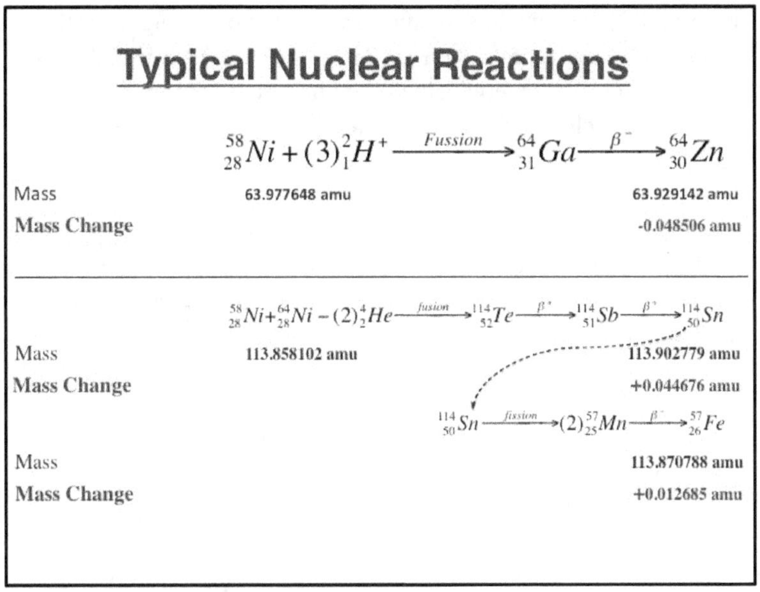

filled, that causes the nuclear reactions? Or, is it the radiation temperature that sponsors fusion/fission events?

One consequence of associating the energy requirement for the nuclear reactions with the blackbody spectra is that all reactions that can occur, do. This is because the entire continuum of spectral energies is available in the far-from-equilibrium blackbody spectra. At the end of the next section, we will see how the 'next', nuclear reaction is selected from among all possible reactions that are pending at any one moment in time.

## F.  Experimental

Now let's return to the mechanisms of thermonuclear fusion/fission under these conditions. Miley's data from electrolysis of nickel-coated micro-spheres (9) provides a suitable data set for analysis. I have inventoried what I believed are the most likely nuclear reactions occurring in the nickel coated micro-spheres, which I will refer to as the electrode. Those reactions that consistently produced observed transmutation products without producing extraneous isotopes are presented in Appendix B. I have used 'isotope of [element]' data extracted from Wikipedia (23) to complete the energy calculations in these tables.

Candidate nickel-deuterium and electrode impurity-deuterium nuclear reactions are tabulated in the first column. The second and third columns are the initial isotopes formed, and the final stable product of its decay. I initially thought that the reversible portion of the nuclear reaction would extend only to column 2, and that the heat evolved from the experimental apparatus would be that from beta-decay of the initial fusion product to stable isotopes. I also suspected that the isotopes observed in the 'post-experiment' electrode would be all of the decay products of the initial fusion/fission reaction. This worked fairly well as long as I made some other assumptions.

First, I had to allow gaseous products to 'gas out' of the apparatus. This seemed reasonable, but then I had to look at the half lives of gaseous intermediaries, to make the judgment call – is it reasonable to assume that this gaseous product had enough time to 'gas out'? Wherever unstable gaseous products were formed in the initial fusion/fission reaction, the half-life of the isotope is provided in the table so that the reader can assess the opportunity for gassing that product out of the electrode before it reacts further. Stable gaseous products were assumed to gas out of the electrode.

Another source of concern at this point was the feasibility of fusion reactions involving 5, 6, …10 deuteron. Having completed hundreds of decay sequences for all types of possible nuclear reactions, it had become apparent that multiple deuteron reactions were the only way to produce most of the low atomic weight products in Miley's Table. But there were practical problems here. Did the multiple deuteron reactions occur all at one time? Or, do they occur sequentially? There seemed to be a proximity issue in a face centered cubic lattice if the reaction involved, for example, 10 deuterons. Yet this still seemed a preferred route, because sequential deuteron addition produced many short half-life, radioactive isotopes that probably were not available for further deuteron addition. Another concern about multiple deuteron reactions is apparent in the reaction involving $^{58}_{28}Ni$ plus six deuterons, yielding $^{70}_{34}Se \xrightarrow{\beta^+} {}^{70}_{33}As \xrightarrow{\beta^+} {}^{70}_{32}Ge$. But, $^{70}_{32}Ge$ is not measured in the post-experiment electrode. It is 'absent'. Is this because the addition of six deuterons to $^{58}_{28}Ni$ never occurs, or occurs only after the first five additions (, which could be all that occur during this experiment's duration), and is not occurring in sufficient amounts to be measured yet in the experiment. Or perhaps, $^{70}_{32}Ge$ undergoes fission to $(2)^{39}_{17}Cl \uparrow$. This result is ambiguous. There are many reactions involving 6 or more deuterons that yield stable terminal isotopes that Miley did measure. I also saw that although the final fission product, chlorine gas, is a plausible fate for the unwanted $^{70}_{32}Ge$, why doesn't every other final, stable isotope undergo fission.

There are also questions regarding the initial amounts of specific isotopes available for reaction. For example the nickel isotopes in the initial electrode are probably present in the normal isotopic composition $^{58}_{28}Ni$ (68%), $^{60}_{28}Ni$ (26%), $^{61}_{28}Ni$ (1.1%), $^{62}_{28}Ni$ (3.6%), and $^{64}_{28}Ni$ (0.9%), indicating that reactions involving $^{58}_{28}Ni$ are far more likely to produce measurable quantities of fission/fusion products as those involving $^{61}_{28}Ni$ or $^{64}_{28}Ni$. Electrode impurities should react according to this same concentration dependance.

Another issue that crops up in the data analysis presented here is illustrated in the fusion of $^{64}_{28}Ni$ with 3, 5, and 7 deuterons. In each case, there are multiple reaction pathways. Is one path preferred over the other? Why is one of the product isotopes absent ($^{70}_{32}Ge$, $^{78}_{34}Se$) even though it occurs along an overwhelmingly preferred pathway?

I have also looked at the range of fusion reactions between the initial electrode isotopes (i.e. $^{58}_{28}Ni + ^{60}_{28}Ni$, $^{58}_{28}Ni + ^{107}_{47}Ag$, or $^{107}_{47}Ag + ^{68}_{30}Zn$), and also the full range of those fusion reactions, but incorporating one or more deuterons (i.e. $^{58}_{28}Ni + ^{107}_{47}Ag + n(^{2}_{1}H^{+})$ ). These pathways produce large numbers of stable isotope products that Miley did not observe, as well as some that were observed.

These are the kind of questions that have kept me up at night.

Overall the tables show that the reactions producing the lower atomic weight portion of the final electrode composition are: 1) fusion reactions of initial electrode isotopes with one or more deuterons, 2) fission reactions of initial electrode isotopes or 'absent' isotopes, or 3) alpha decays. I have also looked at the same three types of reactions, but involving the products of the initial reactions. This was less productive.

The first three columns in the table summarize my initial analysis of Miley's data. It shows that my methods to this point account for all of the Miley isotopes through $^{75}As$, about half of those in the atomic mass range of 76 through 125, and none of the higher mass isotopes $^{126}Te$ through $^{208}Pb$. I had originally suspected that I would find pairs of reactions that produced a net zero mass change. That is, that there would be no net change in energy content in the initial reaction of the reaction sequences when two or more coupled reactions occurred simultaneously. This would satisfy the reversibility constraint. However, I now realize that net zero energy changes are not required for the reactions to satisfy the reversibility requirement. All that is required is that the blackbody spectra has an absorption and emittance quantum of exactly the same energy as the difference between the initial nuclear reaction in the overall reaction sequence, and the final stable products of that reaction. The blackbody form accomplishes this implicitly. The entire continuum of spectral energies is present by definition. Thus, all reactions that can occur, do occur.

The absence of atomic masses above $^{208}Pb$ and the mass accumulation in $^{206}Pb$, $^{207}Pb$ and $^{208}Pb$ is informative in several respects. These three isotopes are the radioactive decay end products of the uranium, actinium, and thorium series respectively. These are the most likely, if not the only, paths to them. Thus, it is reasonable to conclude that nuclei having mass greater than Pb-208 are produced in the electrode. There is no other plausible explanation of how such neutron heavy, lead products could form in the electrode given the composition of the initial nickel electrode and its impurities. More important is my estimation that there simply aren't enough neutrons in the initial system. Neutron formation appears to be one of the fundamental processes taking place in the cold fusion electrode.

The only way that heavy, trans-lead isotopes can form is by rapid neutron capture in the kind of nucleosynthesis that occurs in supernovae. I am out of my element here. But, if the radiation temperature hypothesis presented in this paper can be given any weight, it is not a great leap to the conclusion that the radiation temperature, $T_R$, of our far-from-equilibrium blackbody spectra could also approach stellar supernovae temperatures. Remember, temperature is a derivative. It increases as the frequency of the exchanged quanta, and also because the exchange occurs more frequently as the frequency increases.

Following this argument further, it is particularly noteworthy that no intermediate, radioactive isotopes of the uranium, actinium, and thorium series are present. Decay along these routes produce unstable intermediates that should be detected in cold fusion experiments. Is it possible that the theorized reversible reaction process short circuits the decay steps to a stable end product, producing only a mass/energy change for the overall reaction? In this way none of the radioactive intermediates or time delays associated with long half-lives occur, and cold fusion proceeds without the messy radioactive signatures of other nuclear processes.

At this point, I was out of ideas and assumptions for explaining reaction products that are absent in Miley's data. Still there were outliers. And the one thing that I known with absolute certainty, is that a law of physics cannot have outliers.

As it turns out, the solution to this dilemma lay in this study's initial premise: the reactions involved in cold fusion processes are thermodynamically reversible. The one thing that all thermodynamically reversible reactions have in common is that their evolution is completely described by the Principle of Least Action. Thus, the one rule governing the selection of specific nuclear isotope products, and the exclusion of others can be summarized.

**Rule 1** - All fusion and fission reactions that can occur are candidates. The one that actually produces a product along any reaction pathway is the reaction sequence that satisfies the Principle of Least Action, and specifically the one that results in the smallest mass/energy change regardless of its sign(+/-).

The result of applying this rule to the reactions in Appendix B is shown in column 4. I have calculated the mass of the reactants in column 1, subtracted from it the mass in the final stable product(s) (column 3), and shown that difference in column 4. Bold type is used to highlight the product selected for by the least action principle. In several cases, I have also shown other possible reaction paths to illustrate that the selected one does indeed have the least energy change. In all cases except one, where absent products formed, or where there were several stable isotope choices, the Least Action Principle selects for an isotope in Miley's Table 3. This is true regardless of the sign associated with the overall energy change. The Appendix B table presents the analysis results in order of increasing reaction energy regardless of the sign of the energy change. I call this model the Least Action Nuclear Process (LANP) Model.

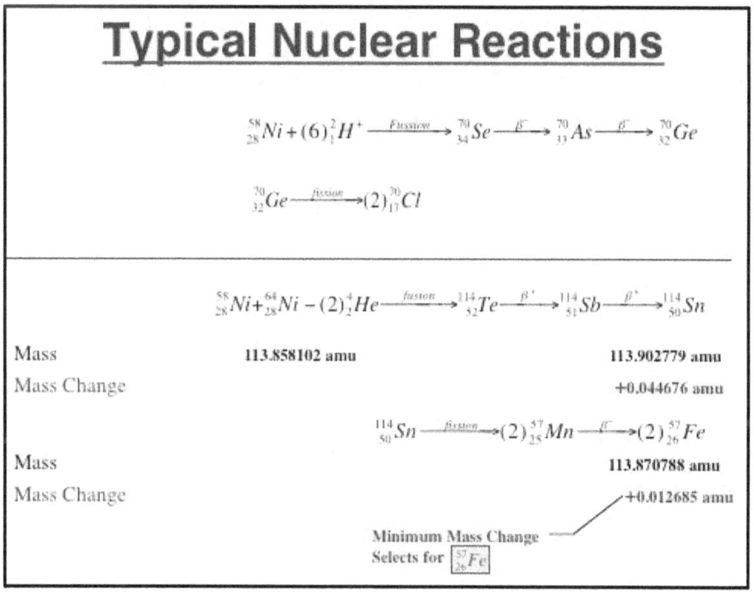


I have tested this model with an independent set of data that was more challenging than the data set used in its development. In particular, I had already drawn about 20 more complex decay diagrams for fusion reactions involving silver and nickel, and also multiple nickel reactants. These produced initial isotopes in the 120-205amu range. In many cases there were more than 10 intermediate decay products, some with extremely long half-lives, and others having low probability decay paths that would normally produce a small fractional of a percent of the total decay product. In all of these cases, the LANP model selects for isotopes in Miley's table, without false positives.

Consider the reaction illustrated in the Figure where the nuclear reaction:

$$^{58}_{28}Ni + ^{107}_{47}Ag + ^{2}_{1}H^{+} \xrightarrow{fusion} \ldots ^{167}_{69}Er \quad \overset{\alpha}{\rightarrow} \quad ^{163}_{66}Dy + ^{4}_{2}He$$

mass change → +0.0775065amu +0.0767937amu

produces 45 intermediate radioactive isotopes and 9 stable isotope products, three of which are in Miley's Table 3: $^{151}_{63}Eu$, $^{155}_{64}Gd$, and $^{163}_{66}Dy$. The results obtained from this reaction sequence show how the Principle of Least Action correctly selects for $^{163}_{66}Dy$, but not along the normal decay pathway shown in the Figure. Instead the Principle of Least Action selects for $^{167}_{68}Er$ with a mass change of +0.0775065amu. This is an end product of the normal decay path. It is followed by alpha decay to $^{163}_{66}Dy$, still within the domain of reversible thermodynamics. The energy change drops

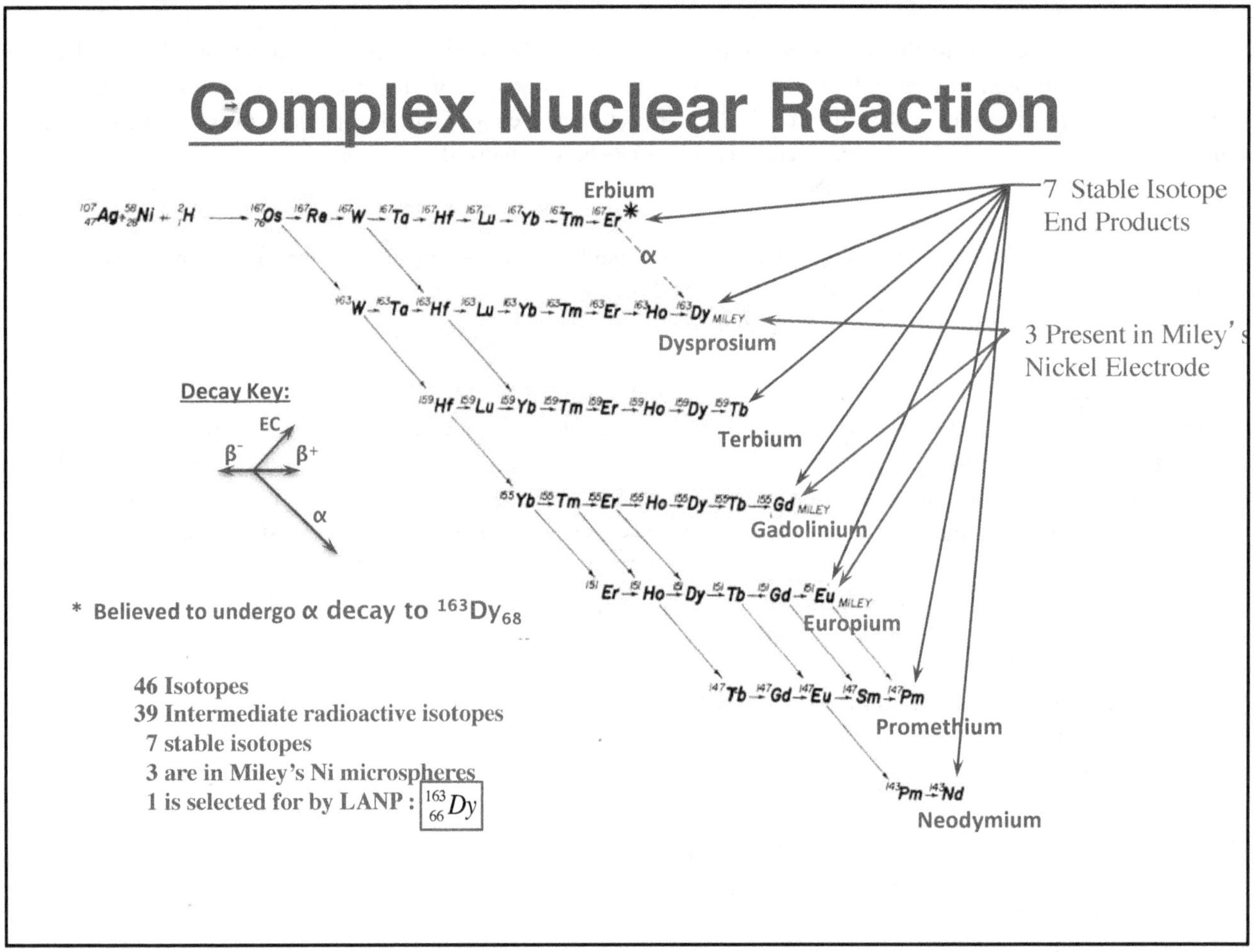


accordingly to +0.0767937amu, the Least Action product. Helium is produced in this final step, but without generating excess heat via the expected pathway (Equation 1). I call this mode of nuclear decay where no radioactive intermediates are formed, and where there are no half-life time delays, $\sigma$-decay.

We are finally ready to look at the issue of excess heat generated in Miley's experiment. The reversibility constraint requires that the overall process be adiabatic. Therefore, we need to explore the limits of that process to identify the step at which it departs from the limiting case of the Second Law where no entropy is produced, and crosses into the domain of irreversibility. First, we note that the mass change appears as a gamma photon in Mossbauer resonance within the far-from-equilibrium blackbody spectra. Negative mass changes increase the total energy within the spectra. Positive changes do the opposite. It seems possible that the summation of these mass changes over all reactions occurring from the point of ignition to the end of the experiment, in some way determines the time history of total excess energy production, or possibly even energy consumption.

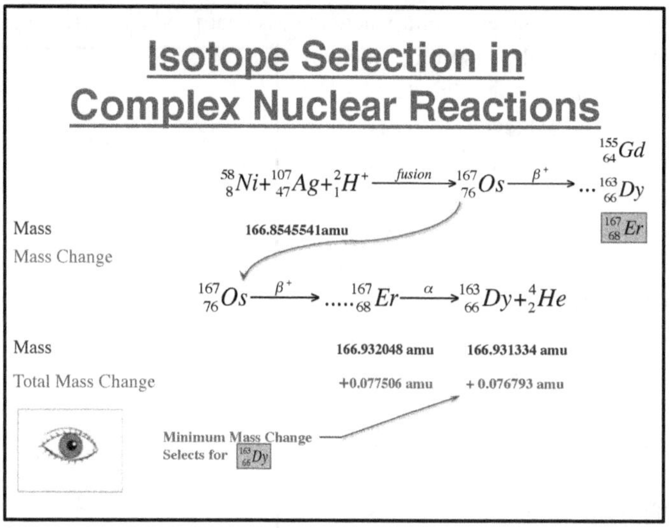


If this energy were to enter the radiation domain at its gamma frequency, but where the number of quanta at that frequency is already saturated, the resulting imbalance necessitates re-distribution into the mass domain. Such an exchange might occur through the Wien frequency channel, where Equation 3 is the transfer function describing the re-distribution of Wien frequency energy into the domain of molecular motion. This increases the thermodynamic temperature, $T_m$, of the experimental device, producing excess heat. Similar thought experiments describe the evolution of endothermic LANP events, and situations in which the blackbody spectrum does, or does not, have vacancies at the gamma frequency.

I believe that such a model will ultimately be capable of not only predicting the amount of excess heat produced, but also predicting the sequencing of nuclear reactions and their time history of endo- and exo-thermic contributions to the heat reservoir. I say this because there are only a finite number of possible nuclear reactions for a given initial isotope mix in the electrode, and if the reaction progression proceeds first with an increment in the Wien frequency, and then energy storage in the radiation domain, or energy re-distribution into the mass domain, it should be possible to very precisely map this process. At each next step, its continued evolution is limited to precisely one nuclear event that satisfies the Least Action Principle. This is a very ordered and exact process.

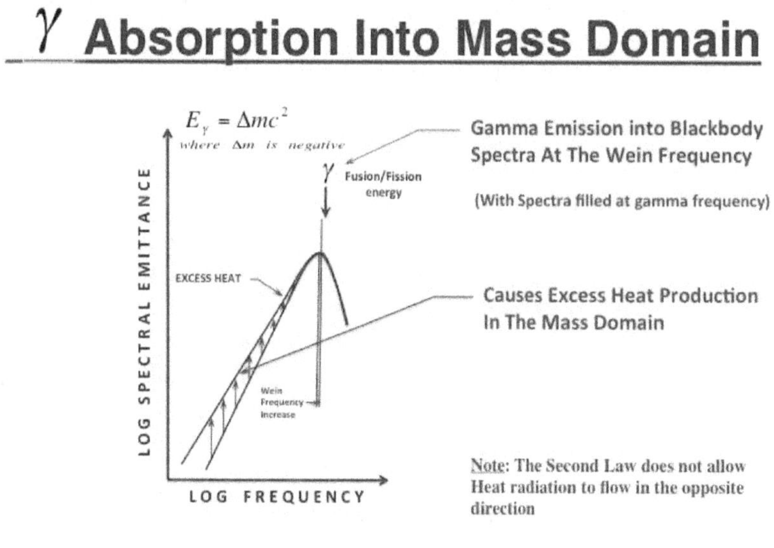

What then is the end point where excess heat production terminates? To place a perspective on this issue, it is necessary to return to the sorting demon discussion in section C. LANP energy is achieved by harvesting random heat of ${}^{2}_{1}H^{+}$ molecular motion, accumulating it in excited nuclear (Mossbauer resonance) states, and then transforming it into thermo-nuclear work. The demon assumes the form of a Ni lattice structure with an affinity for deuterons. By capturing deuterons, and their kinetic energy, he traps heat energy within the lattice, and transforms that random heat of motion into excited electronic and nuclear states, and in this way, decreases the electrode's entropy. But, here is a dilemma. What trick has the demon played on us to first cause deuterons to cascade into the palladium lattice? It is this organizing principle that is his real presence, and the secret of this type of neg-entropic process.

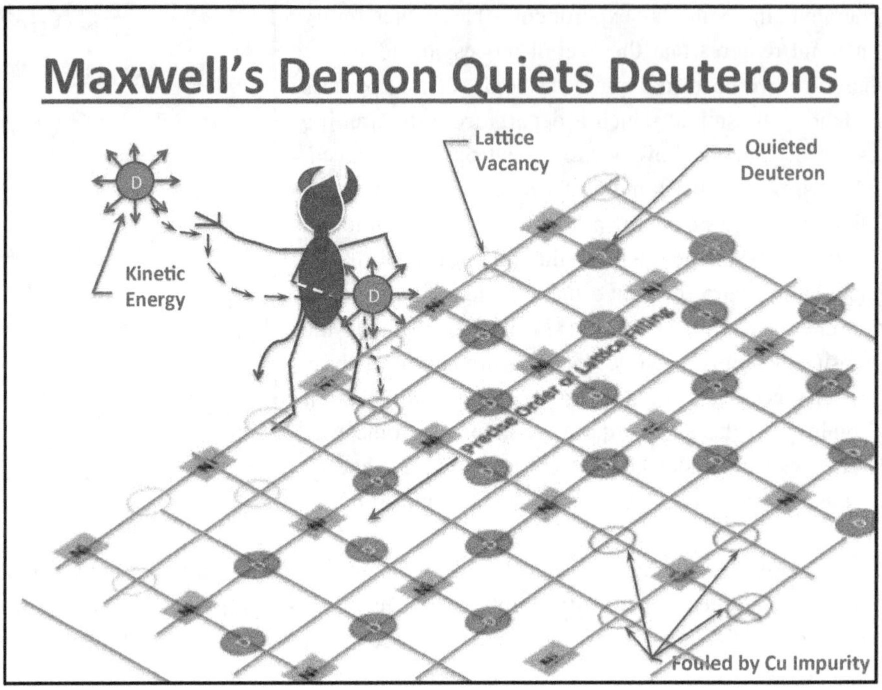


Returning now to the cessation of excess heat production, I can only speculate that isotope transformations distort the ability of the near surface lattice to absorb deuterons. In other words, as the near-surface lattice becomes fouled with atoms that no longer participate in the demons trick, the absorption rate decreases. Nevertheless, deuterons continue to occupy deeper sites, albeit at a slower rate, and nuclear reactions continue until the rate of kinetic energy capture becomes rate limiting.

## G. Some Final Thoughts

The LANP theory is not nickel specific. It can probably be applied equally well to other metal hydrides, and either hydrogen or deuterium absorption. Reactions that can be achieved within the context of reversible thermodynamics occur. Those that do not meet this standard, do not. In fact, there is no reason to limit consideration to metal hydrides. Other processes that are thermodynamically reversible could be subject to a similar theoretical treatment. I would suggest that a good case can also be made that the processes within a living cell might fit into this theoretical framework equally well (24).

The 1-D to 3-D transform function given by Equation 3 appears to be a mathematical statement of the Second Law at the boundary between electrodynamics and mechanics. In its temporal form (19) the equation represents the relative dominance of the forward (entropic) and backward (negentropic) reaction directions. As an example of this function's utility consider its application to the phenomenon of sono-luminescence wherein mechanical energy is converted to electro-magnetic energy. In this case, mechanical energy increases $T_m$ instantaneously without a corresponding increase in $T_R$. Lacking any mechanism to maintain this far-from-equilibrium condition, the system spontaneously moves toward equilibrium by channeling the stored mechanical energy through the Wien frequency channel, and thence, into the radiation domain. If the energy flux's frequency is high enough, visible light is observed.

## H. Discussion

LANP theory is unique in its ability to describe many of the unexplained phenomena occurring in a Fleischmann-Pons electrolytic cell. These include:
A mechanism for loading energy into the metal hydride lattice.
A mechanism for storing that energy until ignition,
A theoretical basis for the fusion temperature requirement and how it is masked,
A mechanism for selecting reactions and products that do and do not occur,
An explanation for the absence of radioactivity.

The theory also has appeal in that it is not nickel specific, or even metal lattice specific, and it provides a plausible mechanism for the solar temperatures that thermonuclear fusion is known to require. LANP is a very hot process.

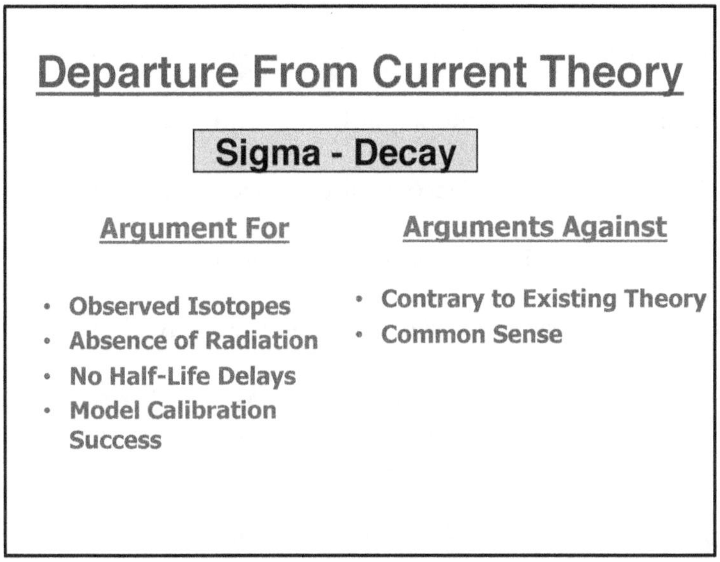
Departure From Current Theory
Sigma - Decay
Argument For
Arguments Against
Observed Isotopes
Absence of Radiation
No Half-Life Delays
Model Calibration Success
Contrary to Existing Theory
Common Sense

The mechanism that causes excess heat to occur requires a detailed methodology for sequencing endo- and exo-thermic reactions, and more discussion of Equation (2), including a more rigorous derivation. Nevertheless, the model demands additional study and experimental work. It answers too many questions to be dismissed.

On the other hand, theoreticians and experimentalists in the field should contain their exuberance for this, or any other promising model. This field is simply too controversial to allow missteps, or premature dialog with the non-scientific community. Places where LANP departs from current theory, and more importantly, from common sense, need immediate study.

For example, is it even plausible that all of the intermediate radioactive decay steps, and half-life constrains can be bypassed by LANP's sigma decay process. The absence of any radiation signature in F&P cells, and the observed transmutation products make that conclusion tantalizing. And yet, it is contrary to everything that we currently know about nuclear processes. The same is true of more fundamental aspect of the theory such as its claim of reversibility. This one feature of the theory is without precedent in modern science, and will be attacked vigorously in peer review. Perpetual motion machines, quite simply, are not supposed to exist. Even more implausible is the claim that stellar and supernova processes might occur within a laboratory device.

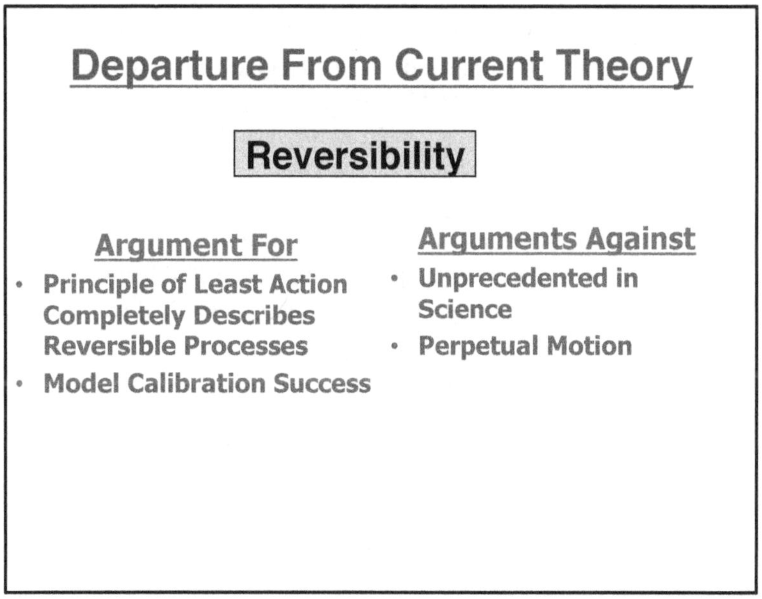
Departure From Current Theory
Reversibility
Argument For
Arguments Against
Principle of Least Action Completely Describes Reversible Processes
Model Calibration Success
Unprecedented in Science
Perpetual Motion

These claims are almost untenable, and yet they seem to constitute a cohesive theoretical framework that is consistent with the data. We should be very careful not to give LANP too much credibility at this point in its short life, and instead design a scientific plan to achieve rigorous experimental proof one way or the other.

## I. References


(1) Fleischmann, M., S. Pons, M. Hawkins, Electrochemically Induced Nuclear Fusion of Deuterium, J Electroanal. Chem., 261, p. 301 and errata in Vol. 263, 1989.

(2) Hagelstein, et al, Input to Theory from Experiment in the Fleischmann-Pons Effect. In ICCF-14 International Conference on Condensed Matter Nuclear Science, Washington, DC., 2008.

(3) Storms, E., Measurements of Excess Heat from a Pons-Fleischmann-type Electrolytic Cell using Palladium Sheet. Fusion Technol.:23 p 230, 1993.

(4) Fleischmann, M., et al., Calorimetry of the Palladium-Deuterium-Heavy Water System. J. Electroanal. Chem., 287: p 293, 1990.

(5) Miles, M., et al., Correlation of Excess Power and Helium Production during $D_2O$ and $H_2O$ electrolysis using Palladium Cathodes, J. Electroanal. Chem., 346: p.99, 1993.

(6) Miles, M. Correlation of Excess Enthalpy and Helium-4 Production: A Review in Tenth International Conference on Cold Fusion, 2003. Cambridge, MA: LENR-CENR.org.

(7) Mosier-Boss, P.A. and S. Szpak, The Pd/(n)H System: Transport Processes and Development of Thermal Instabilities. Nuovo Cimento Soc. Ital. Fis. A, 112: p. 577, 1999.

(8) McKubre, M.C.H. The Need for Triggering in Cold Fusion Reactions. In Tenth International Conference on Cold Fusion. 2003. Cambridge, MA: LENR_CENR.org.

(9) Miley G., J Patterson, "Nuclear Transmutations in thin-Film Nickel Coatings Undergoing Electrolysis", J. New Energy, vol. 1, no. 3, pp. 5-38, 1996.

(10) Karabut, A.B., Kucherov, Y.R. and Sawatimova, I.B., "Frontiers of Cold Fusion" [Proc. 3rd International Conference on Cold Fusion, Oct. 21-25, 1992, Nagoya, Japan], Universal Academy Press, Tokyo, p.165, 1993.

(11) Bockris, J. O'M, Z. Mineviski, Two Zones of Impurities Observed After Prolonged Electrolysis of Deuterium on Palladium, Infinite Energy Magazine, (#5 & #6), p 67, November 1995.

(12) Dufour, J., Murat, D., J. Foos, Experimental observation of Nuclear Reactions in Palladium and Uranium – Possible Explanation by Hydrex Mode, Fusion Technol., 40: p91, 2001.

(13) Mizuno, T., T. Ohmori, and M. Enyo, Isotropic Changes of the Reaction Products Induced by Cathodic Electrolysis in Pd, J. New Energy, 1996. 1(3): p. 31.

(14) Vysotskii, V., Kornilova, A.A., Samoylenko, I.I., Zykov, G.A., Experimental Observations and Study of Controlled Transmutation of Intermediate Mass Isotopes in Growing Biological Cultures, Journal of New Energy, Vol 5, No. 1, pp. 123-128, 2000.

(15) Maxwell, J, C., Theory of Heat, reprinted Dover, New York, 1871.

(16) Boltzmann, L., Lectures in Gas Theory, Translated by Stephen G Brush, University of California Press, Berkeley, 1964

(17) Planck, M., Eight Lectures in Theoretical Physics, 1909, translated by A.P. Wills, Columbia U Press, NY 1915.

(18) Planck, M., Verhandlunger der Deutschen Physikalischen Gesellschaft, 2, 237, (1900), or in English translation: Planck's Original Papers in Quantum Physics, Volume 1 of Classic Papers in Physics, H. Kangro ed., Wiley, New York, 1972.

(19) Szumski, D.S., Theory of Heat I - Non-equilibrium Blackbody Radiation Equation, unpublished manuscript, 2000.

(20) Szilard, L., On the Decrease of Entropy in a Thermodynamic System by the Intervention of Intelligent Beings, translated by Anatol Rapoport and Cechilde Knoller, in B.T. Feld and G.W. Szilard(ed), The Collected Works of Leo Szilard- Scientific Papers, MIT Press, 1972.

(21) Nicolis, G, I. Prigogine, Self-Organization in non-equilibrium Systems, John Wiley and Sons, NY, 1977.

(22) Zhang, Z.L., et al, Loading Ratios (H/Pd or D/Pd) Monitored by the Electrode Potential, Abstracts-ICCF-10, Cambridge, MA, 2003.

(23) Isotopes of Hydrogen In Wikipedia, Retrieved 1/04-7/12, from http://en.wikipedia.org.
(24) Szumski, D.S., Theory of Heat II - A Model of Cell Structure and Function, unpublished manuscript, 2003.

# Chapter 4

# Rethinking Cold Fusion Physics

Published in Infinite Energy Magazine, Issue 120,March/April, 2015

Cold fusion theory is at a critical juncture. Technological development of commercially viable units has outstripped theoretical understanding. It now appears that commercialization will proceed without patent protection, thereby slowing our energy revolution in its finest moment. And as you know, patent protection is being denied simply because we cannot tell the US Patent Office or the PTO how our cold fusion devices work.

What do we do?

### A. A Practical View of the Cold Fusion Process

The theorist's 25-year record suggests that a very different approach is needed. Existing science appears to have exhausted its relevance, leaving us no choice but to consider new theoretical foundations for the cold fusion model. Just how far we may have to depart from existing theory in our quest for alternative science is unknown, but I believe that the path must be one where we 'fill in' the existing science, most importantly, regarding our theory of heat. Specifically, the two existing theories: that for the heat of molecular motion in an ideal gas, and that for the spectral distribution of radiant energy emittance, are limited to equilibrium conditions, and are not interconnected in any single theoretical framework. And yet, the simplest thought experiments make it clear that these processes are connected because of the way that non-equilibrium heat conditions evolve. What are we missing? This seems like a productive direction for discovering the new science that we need. After all, we are dealing with an unknown heat process.

The issue that has preoccupied our attention for 25 years is the coulomb barrier, and in particular, how it is overcome. This has become a Gordian knot, obstructing theoretical discourse unless the theorist can first unravel it. At times it becomes the sole condition for any further discussion. We have to get beyond this. Personally, I find it very satisfying to simply adopt Gottfried Leibniz's principle *of sufficient reason.* Most broadly stated this principle holds that if the coulomb barrier is somehow overcome (and this does seem to be the case in our experiments), then there is a sufficient explanation (albeit unknown) for why this occurs. We need to get on with the more important issues; finding a candidate theoretical framework. The coulomb barrier will fall into place when we find the proper theoretical framework.

Let's begin by taking a very practical view of where we are 25 years later. We continue to deny that we are talking about a perpetual motion machine. However, if we were to stand by our gut feelings, even this would be a modest boast. Perpetual motion only produces as much energy as it takes to operate it. Here, we appear to be producing far more energy than we input. Let's begin by agreeing that it is not magic, or a process with teleological overtones. Let us also agree that it is a heat process; a very far-from-equilibrium one. If indeed it is perpetual motion, or some variant thereof, the only statement that we need to make about it is that it has to operate at the very limit of the Second Law of Thermodynamics where all processes are thermodynamically reversible. To stop anywhere short of this limit leaves us within the domain of irreversible thermodynamics, the comfort zone of most physicists, but the domain where physical theory seems to have exhausted its relevance.

The equilibrium nature of the existing heat theories is unsatisfying and particularly limiting. By any measure that I know, the cold fusion process is anything but an equilibrium state. Indeed, if we look at it from an entropy standpoint, it seems to be best represented as a very low entropy condition, and consequently, one that is very, very far-from-equilibrium. Here again, the underlying physics can best be described as a reversible thermodynamic state. I respectfully submit that we should try using this alternative thermodynamic framework as we move forward.

The physicist in all of us bristles at this idea. Max Planck, the theorist responsible for our equilibrium heat radiation theory, is often quoted in this regard: *"Reversible processes have, however, the disadvantage that singly and collectively they are only ideal: in actual nature there is no such thing as a reversible process."(Eight Lectures in Theoretical Physics, 1915)* Yet, Planck cites several reversible processes in his lectures, and goes on to state that in the physics of the future the most important division of all physical theory will be into reversible and irreversible processes, with the distinguishing characteristic: "In the differential equations of reversible processes the time differential enters only as an even power, corresponding to the circumstance that the sign of time can be reversed." (ibid, Planck, pg. 19). These are then, processes that are equally valid in both the positive and negative time directions, or more accurately, processes that teeter at an apex where movement in either time direction is equally probable. In fact, the theory says that at any point in time, the *next step* in any reversible process can be completely specified by the Principle of Least Action. [Planck's Lectures should become our reference when discussing reversible thermodynamic processes.] But now I am getting ahead of myself.

I will conclude this introduction with my observation that nature will exploit all legitimate scientific nooks and crannies. That's the way I view reversible thermodynamic processes, a 'secret' nook where nature can hide its most marvelous works from our understanding simply because we 'know' that reversible process do not exist in nature. It is ironic that one of the greatest theoretician in classical physical theory may be the person who impedes our progress toward our quest for the unification of physics.

## B. The Reversible Thermodynamic Process

Let's take a minute to look at the reversible thermodynamic state, but through the lens of another Max Planck quote. *"When you change the way you look at things,the things you look at change."* Our argument begins in the domain of irreversible thermodynamic processes where there is always some loss of energy into the domain of random molecular motion. In electrical systems this manifests as dielectric loss to heat of motion. In chemical systems it's the transformations to a higher entropy state and the loss of free energy to heat of motion. In gravitational systems it is the irreversible conversion of potential energy to kinetic energy (motion). And in a photon system, it is the partial conversion of photon energy to heat of motion.

All irreversible processes exist in two time dimensions: the forward direction that we experience in our everyday lives, and where entropy always increases; and the opposite time direction which has the effect of reducing the rate of forward time progression. Consider for example, the transformation of heat from its radiation domain (electro-magnetic energy), to its heat of molecular motion domain. This is the forward direction of energy transformation because it is the direction in which the entropy increases. Heat goes from the more ordered electro-magnetic radiation state to the unordered state of random molecular motion.

It is however, possible to move this heat system in the opposite, or backward time direction by adjusting the boundary conditions to favor a negative time displacement: from the domain of molecular motion, to the radiation domain. This can be accomplished most simply by rubbing a block of wood on a rough surface thereby creating friction generated heat of motion, some of which then partitions to the more ordered domain of heat radiation as the block achieves its new equilibrium temperature. It is particularly noteworthy that regardless of how the heat is generated to arrive at a new, higher equilibrium temperature, be it friction or exposure to a radiant heat source, the final equilibrium condition is identical, and there is no way to determine by which route (forward or backward time displacement) the heating took place.

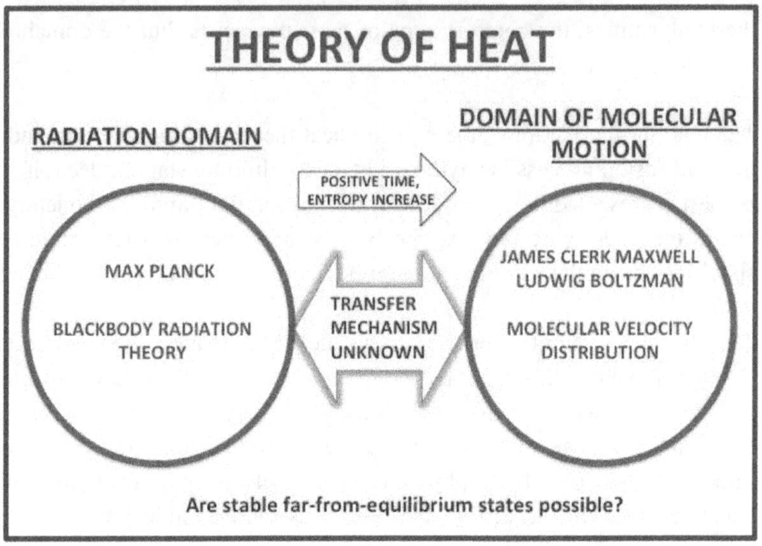


What we have done in this example, is to move a portion of the input energy in the negative time direction, from its less ordered to its more ordered state. In effect, we have changed the relative magnitude of the forward and backward transformations. Another example will make this clearer.

Let's choose as an example, a chemical event, and in particular, the redox event: $2[H^+] + O_2 + 2\{e^-\} \overset{v_f}{\rightarrow} H_2O_2$, where the forward velocity of the reaction is $v_f = k_f[H^+]^2[O_2]\{e^-\}^2$, and $k_f$ has the units per time, and $v_f$, the forward velocity of the reaction, is measured in $moles^5/time$. However, this is only half of the overall reaction:

$$2[H^+] + [O_2] + 2\{e^-\} + e^{H'_f/k_bT} \underset{v_b}{\overset{v_f}{\longleftrightarrow}} [H_2O_2] + e^{H''_f/k_bT}$$

$$2[H^+] + [O_2] + 2\{e^-\} \underset{v_b}{\overset{v_f}{\longleftrightarrow}} [H_2O_2] + e^{\Delta H_f/k_bT}$$

where: $v_b = k_b[H_2O_2][\exp(\Delta H_f')]$

and $K^o_{eq} = {}^{v_f}/_{v_b} = \frac{k_b[H_2O_2][\exp(\Delta H'_f]}{k_f[H^+]^2[O_2]\{e^-\}^2}$, the true equilibrium constant for the reaction.

If we then set: $v_b = v_f$, $K^o_{eq} = 1.0$, and the apparent equilibrium constant, $K_{eq}$:

$$K_{eq} = {}^{k_f[e^{-\Delta H_f/k_bT}]}/_{k_b} = \frac{[H_2O_2]}{[H^+]^2[O_2]\{e^-\}^2}, \log K_{eq} = 23.1, E^o_H = 0.68\ volts$$

The positive value of $\log K_{eq}$ indicates a reaction that is heavily favored in the forward time direction. However, while the forward reaction may dominate, there is a small, but still significant reaction in what we would see as the negative time direction. In other words, we see a chemical event that results in $H_2O_2$ production in our positive time direction. But in looking more closely at the process, this is the net event, where a portion of its overall structure occurs in negative time, or the direction of entropy decrease. This hydrogen peroxide formation reaction is purely entropic. We will call it, the reaction's *normal entropic process*.

In an idealized world, it should be possible to alter the reaction to make the backward direction more dominant. This can be accomplished by altering either, or both velocities to make the difference between them smaller. This slows the overall reaction, and effectively decreases its entropy production. Schrodinger would say (*What is Life?, The Physical Aspects of the Living Cell*, 1946) that we have made the process move in the negative entropy direction, or more simply, that it has become a *negentropic process*, a process where the net velocity is positive, but smaller than that of the *normal entropic process*.

Now look what happens when we take this negentropic process to its limit; allowing the forward and backward reaction rates to become identical. This special case, places the process at the very limit of what the Second Law allows. Entropy production ceases, and the process is precisely balanced among all of its possible outcomes. There exists an equal probability of evolving in the forward and backward time dimensions, and all mass/energy and energy conversion outcomes that are thermodynamically feasible, are possible. Time appears to stand still. We call this limiting case of the negentropic process, a *reversible process*. To an observer the process appears to have taken a huge negative entropy step, and is by any measure very far from equilibrium. Does this sound familiar?

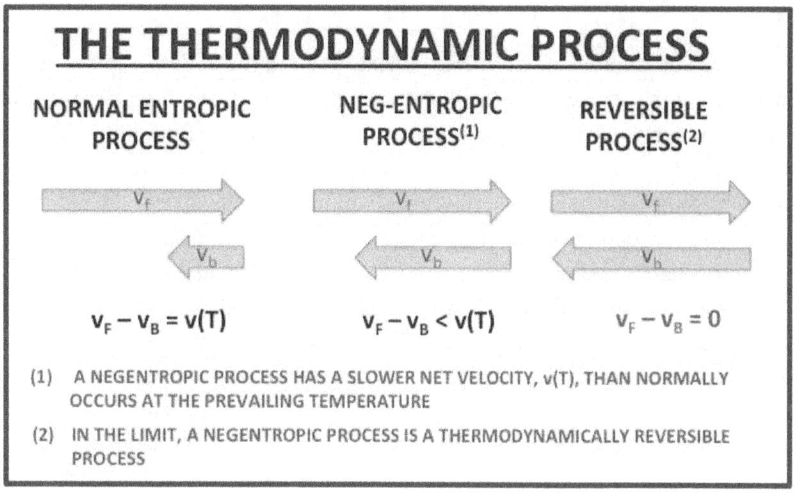


However, the importance of the reversible process in our quest for a cold fusion model lies beyond these circumstances. This is because all reversible processes, regardless of their nature, be they mechanical, electro-dynamical, chemical, or electromagnetic, have one additional constraint that bestows on them their very special place in the sciences. It is the Principle of Least Action. And what makes it so indispensible is the precision with which it specifies from among all of the possible 'next steps' that the reversible process could evolve to, the one 'next step' that results in the *least action*. And if the process remains in its reversible state after this step, there is again only one 'next step' that the process can evolve to, and it too, is determined by the Principle of Least Action. In this way, we see how any process that remains in a reversible thermodynamic state, is firstly *stepwise*, and secondly, must trace out a very specific temporal evolution that we might refer to as the Least Action Process. It does not matter that the process' forcing functions are deterministic or stochastic. As long as the overall process evolves within the framework of thermodynamic reversibility, every 'next step' is precisely determined by the Least Action Principle, and at least in theory, its complete temporal evolution is, in a sense, deterministic. With regard to the process being stepwise, this is a huge benefit to the Least Action Nuclear Process. It cannot cascade in the way that normal nuclear processes do, thereby eliminating the possibility of an explosive event.

Many contemporary physicists consider reversible processes to be rare and not very important in the real world. However, when we consider our palladium or nickel cathode to be in a wholly reversible thermodynamic state, we add a nuance of profound importance that brings a precision and exactitude to cold fusion theory that rivals that which we normally reserve to more traditional areas of physics.

### C. How Does This Apply to Cold Fusion?

The above-described treatment is allowed by the Second Law. But, it is not immediately apparent how net reaction velocities approaching this zero limit are achieved. And yet, this awkward result where the forward and backward progression of time become exactly equal, and where time appears to stand still, is effectively what I believe happens in cold fusion electrodes. It is what gives cold fusion its very far-from-equilibrium character, and its equally far-from-equilibrium product: excess heat.

Now let's return to the question that brought us here: How do our cold fusion devices make the forward and backward reaction velocities exactly equal without violating the Second Law? We begin by affirming that the velocities and time scales that we have been talking about thus far are the averages, or most probable values resulting from large ensembles of events. The laws governing such processes are statistical in nature, and are described by the laws of statistical thermodynamics.

Now let's do another thought experiment, and this time limit our inquiry to the reversible thermodynamic absorption of deuterons at the surface of a nickel cathode. If this is to be a reversible thermodynamic absorption, none of the energy in the reaction space can be energy of random motion. Random motion, a statistical quantity, is foreign to the reversible state, where all events are exactly deterministic. This provides a means for describing our reversible thermodynamic reaction space. In particular, the deuterons in the reactor's heavy water phase are highly mobile, having kinetic energy, $KE_{deuteron} = 1/2\, m_d v^2$, while the massive nickel electrode has zero kinetic energy. Absorption of a deuteron, regardless of the mechanism involved in metal hydride formation, requires that the kinetic motion of the deuteron be 'quieted'...but not lost. The First Law requires that the kinetic energy is conserved on absorption, and we might conjecture that this energy is passed to the nickel hydride lattice where it is stored in a reversible thermodynamic way (i.e. without recourse to post-absorption motion). Ultimately, this energy storage will reside as Mossbauer resonance between identical nuclei, another reversible thermodynamic process.

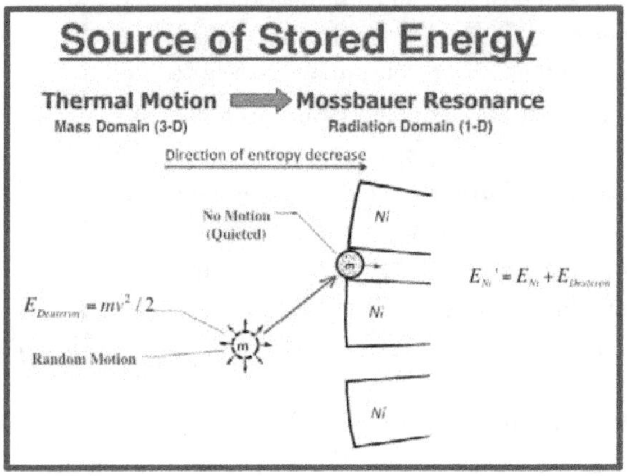


Under these circumstances, immobilized deuterium exists *without any loss of free energy to 'energy of motion'*, and we understand this reaction to be in its thermodynamically reversible state. It has identical probability of going in the entropic and neg-entropic time directions, and thus, its forward and backward reaction rates are identical. In essence, we have taken the nickel hydride formation reaction out of the domain of statistical thermodynamics where the kinetic energy of constituent parts contributes to uncertainty, and placed it entirely within the very precise realm of determinism, where the Principle of Least Action alone determines the overall process' next step.

This state, where the forward and reverse directions of time passage are equal, places us at the very limit of the Second Law, where processes are thermodynamically reversible, energy change is identically zero, and the *apparent* entropy change is negative, even though entropy production has gone to zero. Relative to the rest of nature, this process has taken a step backward in time. This is how cold fusion achieves neg-entropic character in a world of statistical processes.

My next essay will speak to the mechanisms that maintain a very far-from-equilibrium energy storage between nuclei as Mossbauer Resonance, and describe how this energy accumulation achieves solar core temperatures in a cold fusion electrode.

## D. Try This Exercise

My recent paper in the Journal of Condensed Matter Physics presents a cold fusion model based on reversible thermodynamic principles and what I call the Least Action Nuclear Process (LANP). But more to the point, it presents a calibration of the LANP model against nuclear transmutation data from Miley's nickel microsphere experiments. The result show how the model successfully predicts transmutation products in 210 nuclear reactions, and it does so without false positives. In other words, it does not predict anything that is not in Miley's final electrode.

Having spent hundreds of hours doing these calculations with a calculator, I understand the difficulty in replicating my calibration, or the daunting task in repeating them for another electrode (the calculations are electrode specific). Therefore, I have placed all of the relevant calculation sheets on the web so that the workings of the Least Action Principle can be verified for these nuclear reactions. You might want to go through a few calculations to see that it is the Principle of Least Action that correctly specifies the final transmutation products in Miley's data. Table 10 in the web site that I talk about below, is a good place to see this.

My point in providing the data on-line is to show how transmutation products are unambiguously selected for by the Least Action Principle. It is not a very long stretch from that point, to an understanding that this can only occur if the underlying process is thermodynamically reversible. The calculations also show how time delays due to long half-lives disappear, as do the concerns regarding multiple deuteron involvement in any reaction.

The model holds promise in several other important regards.

1. It explains the mechanisms involved in experiments where no excess heat is measured.
2. It provides a mechanism consistent with the laws of physics regarding the absence of radiation or radioactive products in the experiments.
3. It provides, at least a plausible, if unproven, mechanism to explain the absence of a coulomb barrier to nuclear fusion.
4. It shows how both fission and fusion products are selected for by the Principle of Least Action.
5. It provides a rationale for thermonuclear temperatures in a laboratory temperature device.

You might at least go through a few of the more improbable nuclear reactions (i.e. those that produce anomalous isotope distributions, or fission products, or those involving multiple deuterons) to convince yourself that it is indeed the Least Action Principle that unambiguously selects for these products. The data and analysis can be found at **www.LeastActionNuclearProcess.com**.

Finally, Let me refer you to Planck's final word on this subject (A Survey of Physical Theory, 1925), and in particular his discussion of how the Principle of Least Action involves only two quantities: energy and time. The distance relationships, and in particular those involved in the coulomb repulsion, play no role in thermodynamically reversible processes. As long as the next step in the reversible process is thermodynamically possible, it is a candidate, subject only to the Least Action constraint.

# Chapter 5

## Cold Fusion and the First Law of Thermodynamics

Published in Infinite Energy Magazine, Issue 123, September/October, 2015

Low energy nuclear reactions! That doesn't sound quite right. Nuclear reactions are high-energy events that produce lots of heat. The idea that *small* quantities of energy can facilitate these reactions flies in the face of reason. How can you possibly operate a low temperature machine to produce nuclear fusion?

And yet, as certain as we are that this cannot happen, we have memorialized this concept in the names we have chosen for our very special process: Low Energy Nuclear Reaction, Chemically Assisted Nuclear Reaction. Why not simply advertise it as the only nuclear process that allows an end run around the First Law of Thermodynamics.

My first essay spoke to the need for a fresh look at existing science, or perhaps even new elements of science as we move toward a cold fusion theory. I suggested that we consider a modeling framework based in reversible thermodynamics. After all, we are dealing with a very far-from-equilibrium thermodynamic state, and that is precisely where reversible processes take place...very far from equilibrium, and at the very limit of the Second Law. This is in contrast to the current paradigm in cold fusion circles that hopes to discover an existing or new physical process that will initiate nuclear fusion reactions at room temperature. Needless to say, this approach has not produced a viable modeling framework. Nor will it. What is needed, is an expanded view of what is possible... but always within the context of the First Law of Thermodynamics.

### A. Where Does the Energy Come From?

The First Law of Thermodynamics tells us that mass/energy-*in* must equal mass/energy-*out*. So if we are going to have high energy output, we need to either 1) input at least the quantity of energy required for ignition, or 2) somehow accumulate the ignition energy over time. I think that the record is clear on this point. No known physical or chemical process can produce the required ignition energy without reservoir storage. And while it is possible that there is some unknown heat process that might release the required energy at room temperature, I don't think so. Too many of our very learned colleagues have traveled this road without success.

Let's look at the alternative. Is it possible to accumulate energy over the course of, for instance, the loading period, and store it until it is sufficient for ignition? I believe that it is. However, if we are going to explore this avenue, our immediate problem is finding a storage mechanism that obscures this energy until the moment of ignition. We know that it is there. The First Law tells us that. We just need to find out where it resides.

We can identify four types of energy in our cold fusion reactor:

1. chemical bond energy,
2. blackbody electromagnetic energy,
3. electrical energy input for electrolysis, and
4. energy of thermal motion.

Chemical free energy is contained within the subset consisting of the first two, and this is really the only possible place for energy storage in a form that might later be converted to nuclear work. The electrical current is a transient condition. It is not suitable for energy storage. And finally, there is thermal motion, which is simply the most primitive energy form.

However, I don't want to minimize the thermal motion portion of the energy inventory in any way, because as we will see in what is to follow, it could be the ultimate source of energy for the cold fusion process. In that regard, there is one important understanding that we have to grasp regarding thermal motion. It revolves around Helmholtz's great contribution to physics, summarized here by his student, Max Planck: *"In accordance with Helmholtz, heat energy is reduced to motion, and this certainly indicates an advance which is to be placed, perhaps exactly on the same footing as the advance which is involved in the consideration of light waves as electromagnetic waves."(Eight Lectures in Theoretical Physics, 1915, Pg 106).*

The Helmholtz free energy includes that due to motion. The Gibbs form does not. This is the approximation that Gibbs had to make to achieve the simplicity of his physical chemistry methods. In the present case, we will adopt the method of Helmholtz because it is the only complete representation of the cyclic heat processes, and the only method that is suitable to the reversible thermodynamic processes that we will be considering.

My recent paper in J. Cond. Matter Nuclear Sc. describes an energy harvesting process wherein the thermal motion of deuterons in a F&P device is reduced to exactly zero at the first instant when it is absorbed into the metal hydride lattice. The deuteron's kinetic energy goes to zero in this process, but the First Law requires that this energy is not lost, and is instead absorbed into the metal hydride lattice where it is stored as lattice bond energy. I say that the deuteron's kinetic energy has been 'quieted'.

At first glance this appears to violate the second law of Thermodynamics because the process's evolution decreases the system's entropy. But that is a naïve assessment. There is nothing that precludes the transference of thermal motion to electromagnetic energy. I show ample evidence for this in my thought experiments of two identical blocks that are heated to the same initial heat content by friction and radiant heating. But, what is unusual about deuteron 'quieting' is the capture of the electromagnetic energy in covalent bonds or Mossbauer resonance. These are stable far-from-equilibrium conditions that once achieved, prevent the energy's return to the domain of thermal motion.

By this means, we have *" exploited an instability in the normal thermodynamics, causing a branch to what would otherwise be an inaccessible thermodynamic state" (Prigogine, I., From Being to Becoming, 1980).* Ilya Prigogine has called this a dissipative structure. In our case, the inaccessible state is nuclear: fusion or fission, or more simply nuclear transmutation. The instability in the normal thermodynamics is, I believe, the movement of energy from the domain of thermal motion into the domain of reversible thermodynamics, where it accumulates as lattice bond energy, but with no path back to thermal motion. The one thing that you will want to remember about the difference between these two thermodynamic states is the absence of any kinetic energy in the reversible process domain.

The first of two dissipative structures in the cold fusion process is that described above wherein deuterons are captured by a nickel or palladium cathode to form metal hydride. The captured energy is stored first as covalent bonds, and later as Mossbauer resonant energy between nuclei.

It is important that you also see how the dissipative structure has operated in this context. The rapidly moving deuteron cleaves to the cathode's metal lattice, instantly ceasing all motion. It has been 'quieted'. This capture removes its energy from the domain of statistical processes and thermal motion. It becomes instead electromagnetic energy between sub-atomic particles. And it is locked there.

But, there is a more important point. Do you see how this energy accumulation is masked from our observation? The absorbed energy quanta exist solely as bond energy between sub-atomic particles. It cannot be detected, in exactly the same way that we do not observe the shared covalent bond energy in diatomic gasses or the bond energies in the nickel lattice. There is no *apparent* energy accumulation, but it is there.

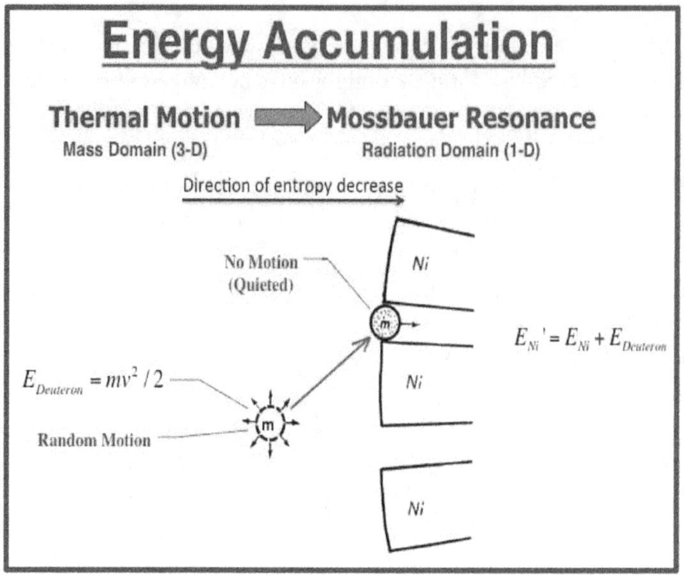


**B. But, Something More is Needed for Ignition!**

While we may have found a way to accumulate the energy required for ignition, as it turns out, the First Law is not a sufficient condition for ignition. We are looking at a *thermodynamic* process with First and Second Law requirements. These are processes that are temperature specific in their operation. And as we all know, the operating temperature for nuclear transmutations is essentially solar core conditions. Now, I know that you are saying to yourself "he's not going to show how solar core temperatures might occur in a laboratory temperature device". Humor me. Let's begin by putting on our eyes of discovery as we immerse ourselves in a fascinating question, and the subject of my research report at ICCF-19: "The Atom's Temperature".

Let me state my findings very briefly.

Temperature is a derivative. It is expressed as joules/sec or joules/$m^2$-sec. At every equilibrium temperature, the total emittance is unique to that temperature, and characterized as the area beneath a blackbody spectral curve in accordance with Planck's theory. Furthermore, the spectral curve is the same in the interior of all materials at that temperature... regardless of its composition, *and no matter how small.*

Let's now take a look at the emittance between two identical nuclei in Mossbauer resonance. You might envision this resonance as the continuous exchange of a gamma photon between the two nuclei. This is known to be a recoilless process, or more accurately a thermodynamically reversible process. There is no kinetic energy present, and the exchange continues indefinitely, without loss of energy to motion, and with no change in entropy.

So, lets take a look at the temperature of this system of two atoms. The emittance of either is the energy of the resonant gamma photon times the frequency of its exchange:

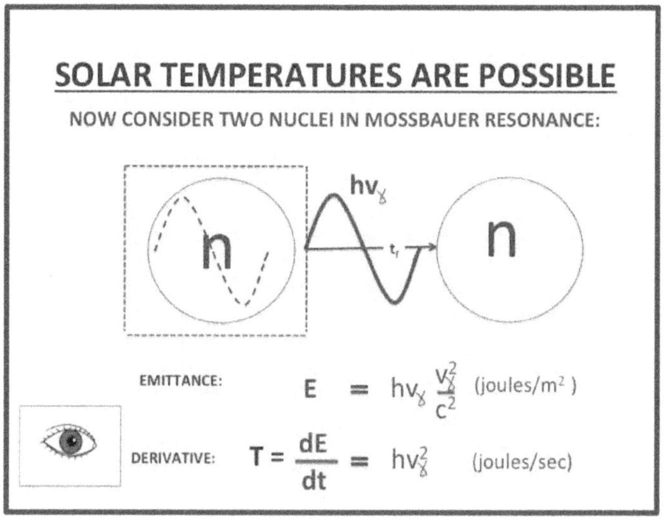

Thus: $$T(^oK) \xrightarrow{proportional} h\nu_\gamma^2$$
where the gamma frequency, $\nu_\gamma$, places the atom at solar core temperature, or above $10^7\ ^oK$.

## C. Gamma Emissions Decay in Accord with the First Law of Thermodynamics

The second dissipative structure occurs when sufficient energy has been absorbed to allow ignition. You can view this as a roller coaster ascent. The chain drive continually lifts the coaster toward the top, and then releases the accumulated potential energy into kinetic energy as the train pulls away from the chain drive. In the case at hand, stored electro-magnetic energy becomes great enough to dissipate into nuclear transmutations, never to re-enter its resonant energy state again.

This is the second place where the First Law is implicated in our understanding of the cold fusion process, and where a slightly different way of looking at the process gives us valuable insight into another conundrum: the absence of gamma radiation in F&P experiments.

Regardless of whether heat is produced or not, all of the nuclear reactions taking place in a Fleischmann-Pons cell, emit or consume relatively large amounts of energy. Because the emitted energy is at the gamma end of the energy spectrum, it should be easy to measure. But as we all know, this is not the case. With rare exceptions, no form of radioactivity is ever observed. So what happens to the energy given off by these nuclear fusion and fission reactions?

Let's first remember, we are dealing with a heat process, one that is different than any other that science has ever encountered. Not only does it produce new stable nuclei, and generate heat in quantities that can only be explained as being of nuclear origin, but it does so without emitting any of the radioactive byproducts that we expect. The first Law of Thermodynamics requires that we either 1) see gamma bursts, or 2) account for the gamma energy in some other way. So let's look at the process through lens number 2.

The result is very satisfying. We find that the LANP process is doing exactly what it is supposed to do. It conserves energy as required by the First Law. In particular, we find total energy production that is consistent with nuclear reactions and gamma emission, but it is simply in the wrong form...low energy heat rather than high energy gamma bursts. The First Law only requires that these large amounts of energy are conserved. It is our task, to determine why it is in the wrong form.

The LANP theory explains this paradox in a straightforward manner. Consider the microwave that you used this morning to reheat your coffee. It emits microwave radiation at 2.45Ghx and imparts that energy to your coffee. But the radiation coming from your coffee is a frequency-transformed representation of the original microwave energy, having *essentially* the same total energy, but a lower energy representation of it. A similar frequency transform occurs in the LANP reactor. The gamma photon is absorbed into the far-from-equilibrium blackbody spectra where it is transformed to lower energy photons that are spread over a large number of lower frequency energy quanta. *The summation of these is the total energy of the original gamma photon*. This is the excess heat that we measure. Its origin is nuclear gamma emissions.

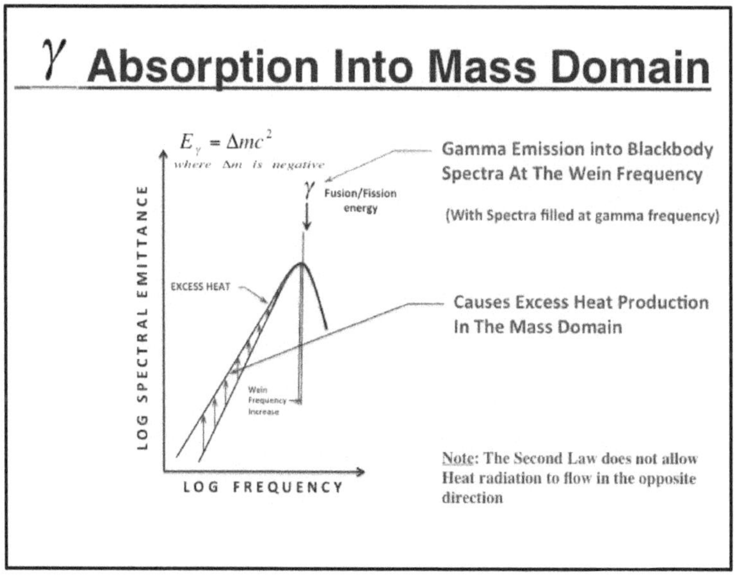


It is also noteworthy that no other radiation, or

radioactive waste products result from LANP operation, nor do we see half life delays. The nuclear transmutations seem to occur directly between isotope reactants with deuterons, and the final stable isotope products. In this Least Action process, there are no traces of unstable intermediate isotopes or their radioactive decay products. Reversible thermodynamic processes are characterized only by their initial and final states, without any of the messy intermediate stuff. All that we can say about the path between the two is that it is precisely and completely characterized by The Principle of Least Action.

# Chapter 6

## Can We Explain Excess Heat Uncertainty with a Law of Physics

Published in Infinite Energy Magazine, Vol. 22, Issue 128, July/August, 2016

The most potent criticism of cold fusion experiments centers on our inability to produce excess heat reliably. The argument goes something like this. Natural laws are exact and reproducible. These experiments are neither. Most critics conclude that the heat effect is due to measurement error. Others, including some cold fusion scientists, say that there must be some other factor, present in some experiments and not in others, that produces excess heat in such a random fashion. Nowhere else in nature, has a law of physics produced seemingly random results like this.

So we have this dilemma. Those of us in the cold fusion field believe that the excess heat effect is real, and that the underlying process is indeed nuclear. But, just like the mainstream physicist, we have to ask: How does a Law of Nature produce such random results, not just for the presence or absence of excess heat, but also its time series? This essay will show how a completely deterministic system, operating according to natural laws, can exhibit this type of random behavior, but only if the underlying process is thermodynamically reversible.

### A. Primary and Secondary Effects

We conduct our experiments as if excess heat was the primary experimental effect. But it is not. Nuclear transmutations occur at a much more fundamental level in the cold fusion process, and are thought to be responsible for the excess heat and low level radiation that are routinely reported. In fact, Tadihiko Mizuno's meticulous experiments show that nuclear transmutations occur even when no excess heat is observed (1). This single observation suggests that the switching mechanism driving the 'excess heat' 'no excess heat' portion of the process is not the presence or absence of nuclear reactions, but possibly, a law of nature that modifies excess heat production, and possibly, the other two anomalous behaviors: the absence of gamma radiation, and the absence of unstable end products.

### B. The Principle of Least Action

Laws of Nature are always precise and reproducible. How can a law of physics produce what appears to be completely random output?

LANP theory (2,3,4,5) provides an answer to this question, only because its process is thermodynamically reversible. These are processes that operate at the very limits of the Second Law where entropy production ceases. Every next step in any, and all, reversible thermodynamic processes is determined precisely and unambiguously by the Principle of Least Action (6). In the LANP model 'least action' is taken as the 'smallest mass/energy change'.

The nuclear transmutations occurring in our metal lattice electrode involve reactions of deuterium, with the base metal

isotopes [ $Ni_{28}^{58}$ , $Ni_{28}^{60}$ , $Ni_{28}^{61}$ , $Ni_{28}^{62}$ , $Ni_{28}^{64}$ ], and the isotopes of impurities in it [eg: $Si_{14}^{28}$ , $Si_{14}^{29}$ , $Si_{14}^{30}$ , $Cr_{24}^{52}$ , $Cr_{24}^{53}$ , $Fe_{26}^{54}$ ,

$Fe_{26}^{56}, Fe_{26}^{57}, Fe_{26}^{58}, Ag_{47}^{107}, Ag_{47}^{109}$, etc]. The nuclear transmutations can be either exothermal, or heat producing with a

positive sign: $${}^{58}_{28}Ni + (1){}^{2}_{1}H^{+} \xrightarrow{fusion} {}_{...28}^{60}Ni \qquad + 0.018658\ amu^{**}$$

or endo-thermal, or heat consuming with a negative sign:

$${}^{58}_{28}Ni + {}^{107}_{47}Ag + {}^{2}_{1}H^{+} \xrightarrow{fusion} ...{}^{163}_{66}Dy \quad - 0.0767037\ amu^{**}$$

Every next step in the Least Action Nuclear Process is that which produces the smallest mass/energy change *regardless of its sign*. All heat exchanges take place through the far-from-equilibrium blackbody spectral distribution in the electrode's covalent and Mossbauer resonant bonds. [** www.LeastActionNuclearProcess.com for details of calculations]

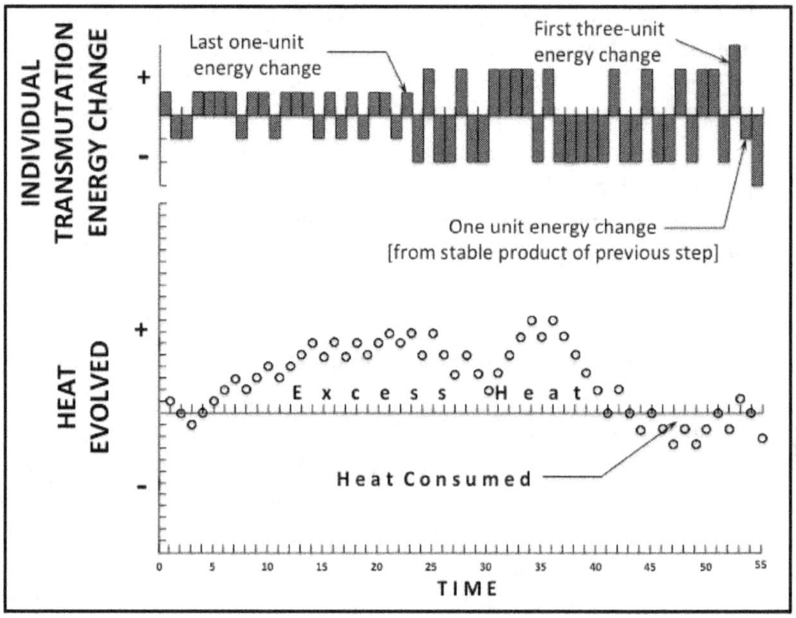


Consider the upper graph in this figure. It displays the energy changes from a *hypothetical* least action process, plotted as a time series. First, I want you to see how the process proceeds in the direction of increasing mass/energy change. I have rounded these energy changes to the nearest one, two or three units to make the lower graph more easily understood. Now notice how the individual transmutation energy changes tend to be positive in the first portion of the graph. This causes a net positive change in the process temperature, and excess heat is observed. Now watch what happens in the last third of the transmutation-energy-change time series where negative energy changes predominate. Heat production decreases...going negative, and no excess heat is observed late in the experiment. I also want you to notice the *one* unit change occurring at the end of the time series. Anomalies like this occur because deuterium reactions with transmutation *products* are continually adding new low energy changes that further distort the transmutation-energy-change time series and consequently the excess heat profile.

Now consider what happens when the electrode composition changes by the addition or subtraction of specific stable isotopes in the initial electrode. The entire Least Action Nuclear Process changes. And although the underlying law of nature is both precise and completely deterministic, a much different excess heat time series is observed. In this way we see how apparent randomness is introduced into the completely deterministic LANP process.

But, the more important point that I want to make, is how this seemingly random response occurs because, and only because, every next step is exactly and unambiguously determined by the Principle of Least Action. This is how laws of nature behave, in accordance with strict determinism. But, this happens only because the underlying process is thermodynamically reversible. I hope that this example makes it clear how randomness and non-reproducibility can occur in cold fusion experiments.

## C. Experimental

Allow me to explain another little known observation related to this discussion. We have seen that it is the mix of metal lattice atoms and impurities that determines the characteristics of the excess heat time series. It then follows that there will be experiments where the negative mass/energy changes will predominate throughout the experiment, and little or no excess heat will be observed. But, the LANP model predicts that nuclear transmutations will have occurred even in these 'no excess heat' experiments. The meticulous experimentalist, Tadahiko Mizuno, sent me his data sheets (1) on just such a 'no excess heat' experiment. His before and after SIMS analyses show a broad range of transmutation products even though no excess heat was measured. Dr. Mizuno completed the post-experiment SIMS analysis when most of us would probably not even consider measuring isotope fractions in an electrode where no excess heat was produced. But remember, heat is the secondary effect. Nuclear transmutations are far more fundamental to the process. This is the genius of Tadahiko Mizuno.

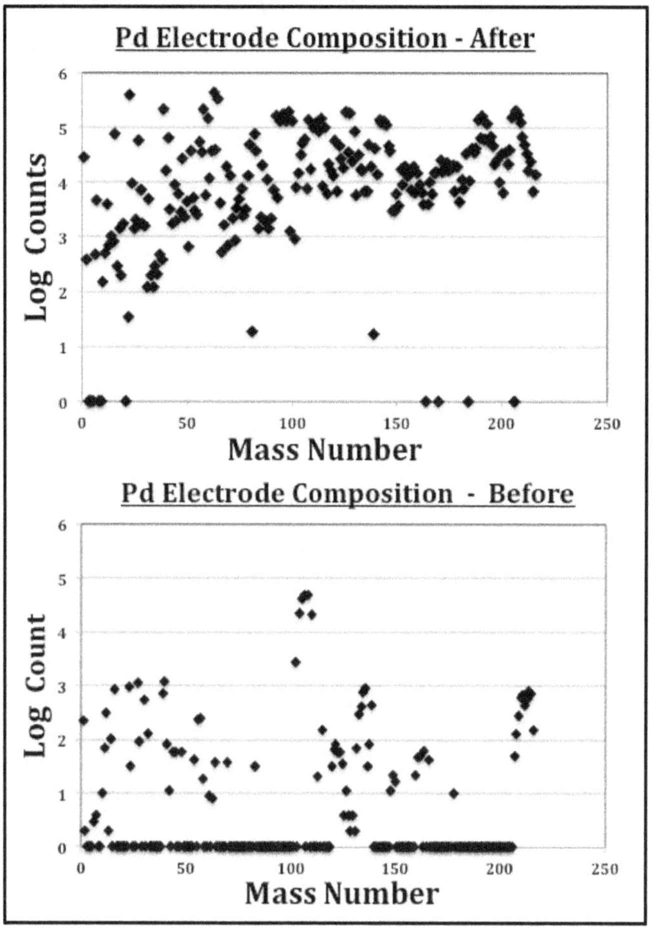


## D. Conclusion

This analysis is informative in the way that it disaggregates 'excess heat' from the more fundamental process in cold fusion experiments: nuclear transmutations. If our goal is to simply commercialize the process, excess heat is a reasonable, primary experimental variable. In fact, if the mechanism between nuclear transmutation and excess heat was unambiguous, this might be a perfectly good way to proceed. However, as we have discovered, this is not the case, and our experiments have not yet found an experimental variable that allows us to modify the heat response in predictable ways.

If we are to understand this process, we have to study it at its more fundamental level: nuclear transmutations. When the nuclear process has a working theory, the mechanism responsible for modifying the heat output of the nuclear process can be studied in a more productive way. One fruitful scientific approach is the LANP theory of cold fusion, which seeks to understand the fundamental process. With this theory in hand, it then becomes possible to show how the primary processes output in modified, according to a law of nature, to produce the randomness that we see in our excess heat experiments. Again, Tadahiko Mizuno was entirely correct in titling his book: Nuclear Transmutations: The Reality of Cold Fusion (7).

## E. References


(1) Mizuno, T. , Experimental data file 'PDbefore.XLS', personal communication.

(2) Szumski, D.S., "Consequences of Partitioning the Photon into its Electrical and Magnetic Vectors upon Absorption by an Electron", In the Nature of Light: What are Photons? V. Chandrasekhar Roychoudhuri; Al F. Kracklauer; Hans De Raedt, editors, Proceedings of SPIE Vol 8832 (SPIE, Bellingham, WA), 883201, 2013.

(3) Szumski, D.S. "Nickel Transmutation and Excess Heat Model Using Reversible Thermodynamics", J. Condensed Matter Nucl. Sci. 13 (2014) 554–564.

(4) Szumski, D., “Rethinking Cold Fusion Physics”, Infinite Energy, vol 20, 120, 2015.
(5) Szumski, D., “Cold Fusion and the First Law of Thermodynamics”, Infinite Energy, vol 20, 122, 2015.
(6) Planck, M., Eight Lectures in Theoretical Physics, 1909, translated by A.P. Wills, Columbia U. Press, NY 1915.
(7) Mizuno, T. , Nuclear Transmutations: The Reality of Cold Fusion, US: Cold Fusion Technology, 1998.

# Chapter 7

# The Atom's Temperature

Presented at ICCF-19, Available at www.LeastActionNuclearProcess.com

## A. Introduction

Temperature is a derivative. By convention (1) it is the rate of radiation emittance from any material body expressed either as J/t, or more practically, $J/m^2$-t. It is uniquely defined by a temperature-specific, equilibrium blackbody spectrum, which is identical in the interior of all material bodies at the same temperature, regardless of their composition, and as we will see, regardless of their size. Max Planck's theory (1) of the blackbody's equilibrium spectra, introduced the quantum into physics.

Temperature also has foundations in molecular motion. It was Helmholtz's great contribution to physics that heat reduces to motion. Planck (2) places this understanding on an equal footing with James Clerk Maxwell's treatment of light as electromagnetic waves. The Maxwell-Boltzmann equation (3,4) for an ideal gas's molecular velocity distribution expresses this theory's fundamentals.

While there has been theoretical consideration of the relationship between these heat theories (see (5)), the only theoretical framework relating the two in a quantitative way, is a model containing two temperatures (5), one for the radiation domain, and another for the domain of molecular motion, or what I will call the mass domain. This non-equilibrium heat theory recognizes heat transfer between the mass and radiation domains, and is elegant in the way that the two temperatures can be separated in a far-from-equilibrium manner. The theory's principle shortcoming is a small, but real departure from Planck's theory when the two temperatures are at equilibrium. It is possible that this difference might itself be a fundamental element of heat theory.

The foundations of temperature as a measure of heat content lie in the probabilistic domain of statistical mechanics. In Planck's radiation form, the equilibrium state's statistics are derived from the energy distribution of harmonic oscillators; while in the mass domain, it is the statistical distribution of molecular velocity in an ideal gas, which is distributed according to the Maxwell distribution (3). Because these laws are statistical, there should be limits to their validity as the number of participating energy units decreases. Here we will consider the limiting case where the statistical foundations of the Second Law transition from probabilistic to wholly deterministic. It is in this limit, that we will enter the domain of the thermodynamically reversible process.

## B. The Photon's Irreversible and Reversible States

Our quest for the atom's temperature begins with an understanding of the photon, and in particular, the two mechanistic pathways accessible for its evolution in the presence of an electron (5). This theory of photon absorption is the first of two elements of new science required by the Least Action Nuclear Process model of cold fusion.

Photon absorption is considered to be a two step process. In the first step, the photon is wholly absorbed into the electron but retains its identity as a reversible process. In the second absorption step, the photon evolves into the electron's three dimensions. The electrical and magnetic components decouple, and the electrical vector evolves to three-dimensional charge (with its dielectric loss), and as a consequence, enters the domain of irreversible thermodynamics. The magnetic vector on the other hand has no three-dimensional equivalent, remains one-dimensional, and contributes to the electrons one-dimensional para-magnetic spin. This asymmetrical evolution from one dimension to three, distorts the space/time fabric, giving rise to the characteristic blackbody emittance spectra.

However, there is nothing requiring that the absorbed photon evolve into three dimensions. Consider the possibility that it is immediately re-emitted in precisely the state that had only undergone that first step of photon absorption. The emitted photon is still 1-D, and remains in its reversible thermodynamic state. But let's go one step further and allow this emitted photon to undergo first step absorption by a second electron...and then be re-emitted still as 1-D, and re-absorbed by the first electron, and so on. We have just arrived at the quantum electro-dynamic description of a covalent bond (6). Its photon energy is locked between two electrons, and remains unchanged indefinitely. This view of the covalent bond allows us to assign it a definite energy, while allowing for a broad range of energy conditions where either of these atoms is covalently paired with a different atom, or in a different molecular structure.

First, note how this resonant bond is a wholly reversible thermodynamic quantity, existing without any losses, and with no change in its entropy condition. The effective distance between these covalent electrons is the shared photon's wavelength, $\lambda$. Secondly, this model describes the permanence of the bond. It is this stability of the covalent bond that gives it the very special place in the energy structure of biological and chemical molecular forms.

## C. Temperature of the Irreversible Thermodynamic Process

The area beneath any of the temperature curves described by Planck's blackbody radiation spectra (Figure 2) uniquely defines a specific temperature:

$$T \propto \int K(\nu)d\nu = \int h\nu \frac{\nu^2}{c^2} \cdot \frac{1}{e^{h\nu/k_bT} - 1} d\nu \quad (1)$$

Where $K(\nu)$ is the blackbody spectral emittance; the factor $\nu^2/c^2$ is the conversion factor: per $m^2$; and the statistical representation in Equation (1) includes both radiation domain heat energy, and heat of molecular motion. The temperature measurement is independent of the material's composition, and as it turns out, it is also independent of its size.

Before going further in our understanding of how this irreversible process definition of temperature might differ from that of the reversible thermodynamic state, I need to divert your attention to a conundrum that arises in the contemporary interpretation of Planck's equilibrium blackbody theory, and which is central to what is to follow. In particular, the range of Planck oscillator energies in Figure 1 is far beyond the normal range for electron absorption and emission. It extends into the X-ray and gamma portions of the spectrum where the oscillator energies can no longer be emitted by, or transferred between, electrons. Meulenberg (7) has pointed out that this spectral continuum has to

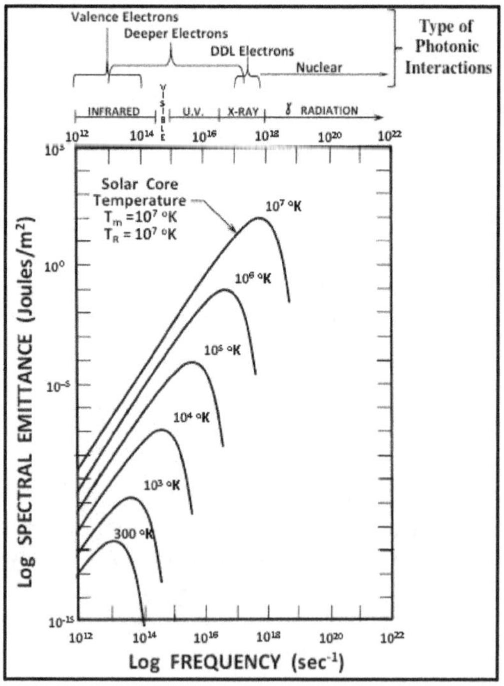


**Figure 1**- Planck's equilibrium blackbody spectral curves, illustrating the types of photonic interaction required of harmonic oscillators across the spectrum.

include intermediate energy exchanges that can only occur between higher energy electrons, probably extending to Deep Dirac Level (DDL) s-shell electrons. Szumski (8) has proposed a further extension of this continuum to Mossbauer resonant nuclear bonds, extending into the gamma range shown in Figure 1, and possibly into the far-gamma range. In these Mossbauer resonant bonds, gamma energy photons are emitted by lattice nuclei, and absorbed by its 'covalent' nuclei, then emitted by that nuclei, and absorbed by the original nuclei, and so on. As in the covalent bond case, this process is thermodynamically reversible, conserves energy, and functions without entropy change. What resonant electron and resonant nuclear bonding might look like over the blackbody spectrum's frequency range is shown at the top of Figure 1.

Do you see how these spectra necessitate an energy continuum from very low energy electron transitions, through what can only be described as nuclear energy transitions, or in this theory's terminology, nuclear resonant bonding. The validity of these curves is well established over this temperature range.

One final point in this regard. My calculations of Widom-Larsen's (9) 'heavy electrons' finds electron masses more than double those of 'standard model' electrons. These turn out to be equivalent to the mass-plus-energy of Meulenberg's (10) DDL electrons, which is a much more satisfying way of looking at the Widom-Larsen model's fundamentals.

## D. Temperature of the Reversible Thermodynamic Process

My purpose here is to discuss the temperature sub-model of the Least Action Nuclear Process theory of cold fusion. The discussion begins by noting that nuclear transmutations are routinely observed in cold fusion experiments. This implies that the cold fusion process probably includes thermonuclear temperatures, and the associated thermal energy. But we see neither temperatures in this range, nor fusion energies in our experiments. This doesn't make sense. These are thermodynamic processes that are temperature and energy dependent. As such, it is essential that we understand how the required thermonuclear temperatures are masked from our observation in these room temperature experiments.

So, how is temperature measured in a reversible thermodynamic process? It has to be fundamentally different than the temperature of an irreversible thermodynamic process, wherein heat energy storage is partitioned between the mass and radiation domains, and the process is of a statistical nature. In the reversible process, there is no kinetic energy, and therefore, no mass domain energy storage. In fact, heat energy, or at least that portion that is accessible from the reversible process reaction space, is limited exclusively to the radiation domain. The motion of material particles, and the statistical uncertainty that it introduces, is foreign to reversible thermodynamics.

We will begin our inquiry by affirming that our focus on a reversible thermodynamic process does not change our definition of temperature. It is still the derivative expressing the total radiant emittance from a mass particle per unit time. Either of the covalent electrons discussed above, exhibit a temperature that is proportional to its emittance, $K(\nu)$, measured as the photon energy, $h\nu$, times the frequency, $\nu$, of the individual emissions, but with the emittance of either electron occupying only one half of the process time. Therefore:

(2) $$T \propto \tfrac{1}{2}K(\nu_i) \cdot \nu_i = \tfrac{1}{2}h\nu_i^2$$

And, for nuclei in Mossbauer resonance:

(3) $$T \propto \tfrac{1}{2}K(\nu_\gamma) \cdot \nu_\gamma = \tfrac{1}{2}h\nu_\gamma^2$$

Equations (2) and (3) are infinitesimal subsets of Equation (1). This treatment of nuclei in Mossbauer resonance is the second element of new science in the Least Action Nuclear Process modeling framework.

There are several important observations in this result. First, while Equations (2) and (3) are precise definitions, *they can be misleading*. The energy quanta involved in these temperature definitions cannot exist without a very structured antecedent spectral distribution that has to be filled before these quanta can exist [see Figure 2]. In this Theory of Heat the distribution of reversible process energy accumulates sequentially as $T_R$ increases, while $T_m$ remains constant. In the illustration, $T_m$ equals 300 °K as $T_R$ increases in quantum increments from 300 °K to $10^7$ °K. There is nothing in this theory's fundamentals which precludes solar core, and even stellar nucleosynthesis temperatures between Mossbauer resonant nuclei. This is indeed, a requirement if the transmutation measurements of Mizuno (11) and Miley (12) are to be believed.

Secondly, the energy difference required to affect equivalent temperatures in a reversible and irreversible process is almost unbelievable. Consider the solar core temperatures for these processes shown in Figure 2. The irreversible thermodynamic process requires five orders of magnitude more energy than the reversible process. Coincidentally, this is the observed difference in energy requirements for fusion in the Tokomak, and in our cold fusion reactors.

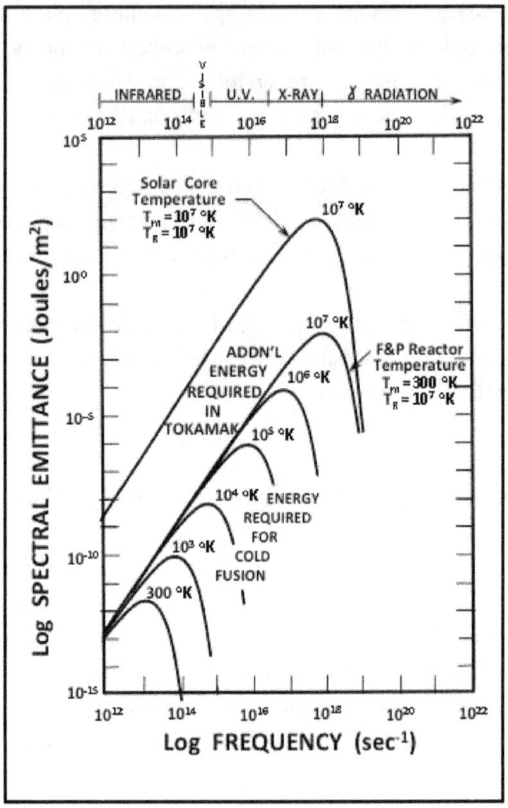


Figure 2- Far-from-equilibrium blackbody spectral distribution for the reversible thermodynamic process in a Fleischmann-Pons cell, wherein all of the energy storage is in the radiation domain. Note the five order of magnitude difference between the cold fusion process and the Tokamak.

Third, we note that this efficiency in the *reversible process* arises only because it bypasses the enormous kinetic energy requirements of the irreversible process. In other words, there is a dynamic equilibrium existing in all irreversible processes, between radiation and mass domain energy storage. In effect, the irreversible process drags around a lot of kinetic energy in order to produce an equivalent free energy (i.e. radiation domain) condition. Cold fusion exploits this advantage.

And finally, the improbable nature of this energy accumulation is dramatically demonstrated when we consider how the radiation domain energy in the metal lattice is accumulated (8). It is the waste heat of random deuterium motion that is harvested (in radio frequency quantum amounts) into the metal hydride lattice, and stored there as resonant bond energy. This flux of chemical (and nuclear) free energy into the lattice creates resonant photons having ever-increasing energy storage. This is the LANP model's forcing function. In effect, the continuous influx of these low energy quanta adds to the far-from-equilibrium energy storage until energies sufficient for nuclear ignition occur. Then energy accumulation continues, establishing ever-increasing process energy, that causes a sequence of stepwise nuclear transmutations, always occurring in an order exactly specified by the Principle of Least Action (5).

## E. Consequences of Thermonuclear Temperatures

It now becomes informative to inquire: what consequence results from this theory of the atom's temperature? Remember how we placed our covalent electrons close enough that they shared a single photon. When we do the same with two nuclei in Mossbauer resonance two things happen. First, the electromagnetic force increases in proportion to the increase in $\nu$, thereby decreasing $\lambda$, and drawing the nuclei closer together. Do you see, in Figure 8, how the electro-magnetic force increases to energy levels that could become operative in overcoming the coulomb barrier in the cold fusion process?

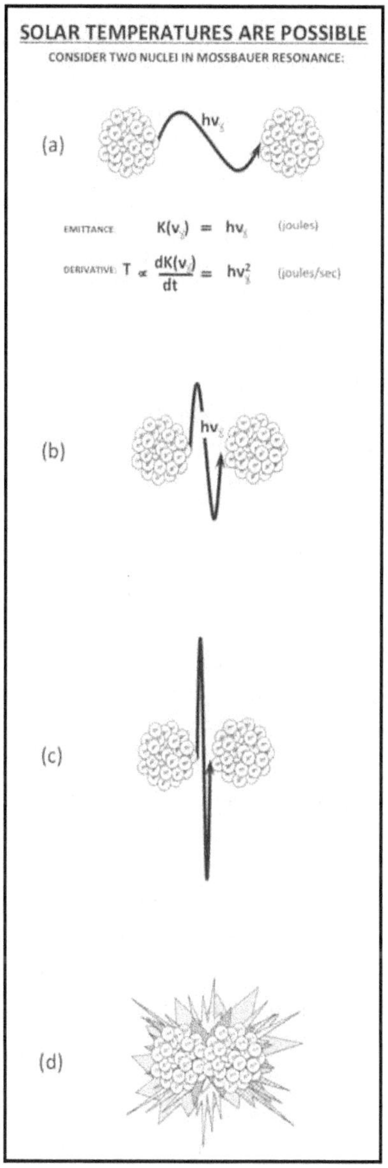


Simultaneously, the inter-nuclei temperature rises in proportion to $\nu^2$, achieving thermonuclear temperatures in a stable far-from-equilibrium manner. As these two conditions converge on impending fusion/fission, this *dissipative structure* "exploits an instability in the normal thermodynamics (an extraordinarily high reservoir of free energy), causing a branch to what would otherwise be an inaccessible thermodynamic state." (nuclear fusion/ fission)[13]. These temperatures must rise to extraordinary levels, if the nuclear products in Miley and Mizuno's data are to be believed.

Finally, I want you to clearly see how both the energy storage for nuclear transmutations, and the lattice temperatures, are localized between sub-atomic particles, and in this way, hidden from our observation. This is consistent with what we see in our experiments. More importantly, it is consistent with the fundamental tenant of physics: nuclear transmutations require both thermonuclear temperatures and thermonuclear energies. Low Energy Nuclear Reactions and similar representations of the cold fusion process are a misnomer. Cold fusion is indeed very hot.

Finally, this model element appears to offer explanations for certain mechanical properties observed in cold fusion experiments. For instance, we might expect that the tightening at the surface of the metal hydride lattice would increase its surface hardness and tensile strength. This appears to be born out by measurement. Furthermore, the surface cracking in electrodes could now be explained as the result of competition between dramatically increased surface tension due to decreasing lattice distances in near surface layers, and the unaffected bulk of the electrode.

## F. Covalent Bonds, and a Lesson from the History of Science

All covalent bonds possess a very specific energy, and according to the theory developed in this paper, exhibit a specific temperature. It is proposed that the same energy relationships could also hold for Mossbauer resonant bonds between nuclei. In both cases, the bond is characterized as a shared photon operating as a reversible thermodynamic process, and at an effective distance of one wavelength. The force that facilitates these bonds is the electro-magnetic force, one of two forces in nature that holds up under careful scrutiny. The other is gravity.

The presence of extraordinary temperatures in these bonds is more speculative. The only evidence lies in the transmutation products measured in post-electrolysis cathodes. That these products could only have formed at nucleosynthesis temperatures is a compelling argument. And, perhaps equally impressive is the compact theoretical package that describes this temperature in relationship to the fusion energy's source, and a plausible solution to the coulomb barrier problem.

This type of speculation, based on theory, is one half of the research process. The other half is experimental inquiry to see if key elements of the theory can be verified by observation. The question then becomes: How can we measure these temperatures and energy quanta? This may be difficult. There is nothing in the literature that I am aware of, where covalent bond strength has been measured directly. The covalent force in biological systems is normally calculated from free energy considerations. Lattice bonding in metals might be calculated as well, possibly from considerations of tensile strength, and the yield properties of the metal. However, direct measurements are preferred. I don't know how this might be done. This is a call to the experimentalists' ingenuity.

We have to also ask ourselves, is it possible that extreme thermodynamic conditions like this might occur elsewhere in nature? Bockris (14) summarized what is generally seen as nuclear transmutation 'fringe literature', including biological transmutation research. More recently, Biberian (15) prepared an exhaustive review of the biological transmutation literature. I believe that it is not a coincidence that biological transmutation research is frequently reported at ICCF conferences. Both cold fusion and the living cell are very far-from-equilibrium processes that continue to confounded scientific inquiry.

Our understanding of biological fundamentals really began in the post-WWII era where experimental discovery seemed to be working hand-in-hand with theoretical advances, and a 'solution' to the living state question seemed at hand. The ensuing 65 years have seen extraordinary advances in our understanding of biomolecules and their systematic place in life. But on the theoretical side, the great promise of a theory of the living state dwindled to the point, where today, only a few individuals pursue Big Picture biology in any serious way. Those few theorists are humbled by the vast chasm between the experimental evidence and our understanding of the living system's fundamental processes. Of particular import is the high degree of variability in biological systems in spite of our intuitive sense that the highly organized and very exact cell structure should be highly deterministic in its function. And yet, it is not apparent that this is the case. There are, undoubtedly, deterministic trends, but these are obscured by a high degree of what is seen as randomness.

The same pattern is emerging in cold fusion research. Here again the variability is large relative to the heat response that we measure. At one end of the experimental response spectrum there is no excess heat evolution. At the other end, there is extraordinarily high excess heat production. At this point in theory development, experimental findings have outstripped theoretical understanding. Like the theoretical biologists before us, we are rapidly exhausting the understanding that contemporary science allows, and theory is floundering.

It is only with the greatest respect for the cold fusion theory community that I suggest, there is one poorly understood area of scientific inquiry that I believe could offer profound insights into both processes. It is reversible thermodynamics; where theory operates at the very limit of the Second Law of Thermodynamics.

Consider the facts. The cold fusion process and the scientific principles underlying the living state are both very, very far-from-equilibrium events. Both seem to defy the Second Law; in the one case by producing large amounts of anomalous heat energy, and in the other, by producing a highly ordered state from light energy and a mostly random landscape of minor molecular forms. Both appear to be operating at the very limits of the Second Law, which is precisely the domain of the reversible process. Is it possible that the fundamentals of both processes might have roots in reversible thermodynamics? Is it possible that the common notion that reversible thermodynamic processes do not exist in nature impedes our ability to understand the fundamentals of both?

## G. References


[1] Planck, M., Eight Lectures in Theoretical Physics, 1909, translated by A.P. Wills, Columbia U. Press, NY 1915.

[2] Planck, M., "Verhandlunger der Deutschen Physikalischen Gesellschaft" vol. 2, 237, 1900, or in English translation: Planck's Original Papers in Quantum Physics, Volume 1 of Classic Papers in Physics, H. Kangro ed., Wiley, New York, 1972.

[3a] Maxwell, J.C. "Illustrations of the Dynamical Theory of Gases. Part I. On the Motion and Collision of Perfectly Elastic Spheres.". Phil. Mag., 4th Series, 19:19-32. 1860.

[3b] Maxwell, J.C. "Illustrations of the Dynamical Theory of Gases. Part II. On the Process of Diffusion of Two or More Kinds of Moving Particles Among One Another". Phil. Mag., 4th Series, 20:21-37. 1860.

[4] Boltzmann, L., Lectures on Gas Theory, Translated by Stephen G Brush, University of California Press, Berkeley, 1964.

[5] Szumski, D.S., "Consequences of Partitioning the Photon into its Electrical and Magnetic Vectors upon Absorption by an Electron", In the Nature of Light: What are Photons? V. Chandrasekhar Roychoudhuri; Al F. Kracklauer; Hans De Raedt, editors, Proceedings of SPIE Vol 8832 (SPIE, Bellingham, WA), 883201, 2013.

[6] Serway, R.A., Vuille, C., College Physics, 4th Ed., Charles Hartford (pub.), 2012

[7] Muelenberg, A., Personal communication.

[8] Szumski, D.S. "Nickel Transmutation and Excess Heat Model Using Reversible Thermodynamics", J. Condensed Matter Nucl. Sci. 13 (2014) 554–564.

[9] Widom, A, Larsen, L, "Ultra Low Momentum Neutron Catalyzed Nuclear Reactions on Metallic Hydride Surfaces", European Physical Journal C – Particles and Fields, Vol. 46(1) (2006), 107-110.

[10] Muelenberg, A., Sinha, K.P., "Deep-electron Orbits in Cold Fusion", J. Condensed Matter Nucl. Sci. 13 (2014) 368–377

[11] Miley, G., J Patterson, "Nuclear Transmutations in thin-Film Nickel Coatings Undergoing Electrolysis", J. New Energy, vol. 1, no. 3, pp. 5-38, 1996.

[12] Mizuno, T., T. Ohmori, and M. Enyo, "Anomalous Isotopic Distribution in Palladium Cathode After Electrolysis", J. New Energy, 1996. 1(2): p. 37.

[13] Prigogine, I., Self-Organization in Non-Equilibrium Systems, Wiley, 1977.

[14] Bockris, J.O.M., Mallove, E.F., "Is the Occurrence of Cold Nuclear Reactions Widespread Throughout Nature?, Infinite Energy, 27, 29-38.

[15] Biberian, J.P., "Biological Transmutations: Historical Perspective", J Condensed Matter Nucl. Sci. 7 (2012) 11–25.

# Chapter 8

# Cold Fusion and the Three Laws of Thermodynamics

Submitted to Infinite Energy Magazine, 1/2/16

## A. Introduction

The three Laws of Thermodynamics are universal. There is no process in nature, either known or unknown, that violates these sacred principles of physics, nor will there ever be one. They apply everywhere, and always; and I sincerely doubt that we will find an exception as we continue our scientific exploration of cold fusion, or its misnomer, LENR.

There is one cold fusion fact that we must be absolutely clear about. Nuclear transmutations require thermonuclear energies and temperatures. Low energy nuclear reactions cannot, and do not exist. For this reason alone, it is incumbent upon theoreticians to focus their creativity on finding the means by which nature allows these energies and temperatures to be present in a way that they remain hidden from our observation. It is my carefully considered opinion, that the coulomb barrier and all of the other anomalous behaviors in our experiments will fall into place if we just keep our attention focused on this one fact about the cold fusion process.

There have been numerous attempts by cold fusion theorists to show how the fusion energy might occur. But in all of those cases, whether it be phonon coherence (1), massive electrons (2), or deep Dirac level electrons (3), the problem is always the same: Where does the process energy come from? And then: How is it stored in a manner that we cannot detect? We are always in this same quandary: the theories seem to require a magic step. This essay evokes the three Laws of Thermodynamics to find the way out of this dilemma.

So, how do the Laws of Thermodynamics produce such uncharacteristic physics? I have endeavored to put this issue into perspective in three other essays published in Infinite Energy Magazine (4,5,6). This is the final essay in that series. It addresses the question: How is the fusion energy required for nuclear transmutations accumulated in a room temperature cold fusion apparatus?

## B. Modes of Energy Storage in the Cold Fusion Electrode

We will begin with the Energy Principle: *Energy can neither be created nor destroyed, it can only change forms*. A thoughtful review finds that there are four types of energy in the cold fusion reactor: 1) Electrical energy in the electrolysis circuit, 2) Chemical free energy between chemical constituents, 3) Blackbody heat radiation, and 4) The kinetic energy of molecular motion. There is in addition, the mass constituents and their implied Einstinian energy equivelance, $\boldsymbol{E} = \boldsymbol{mc^2}$.

These energy components then become initial candidates for the undefined energy storage mechanism. But it is not immediately obvious that any of them is suitable. An increase in the electrical circuit energy would be seen as higher amperage, increasing electrical resistance, and the consequent losses as waste heat to molecular motion. Increased chemical free energy would cause chemical phase redistribution, with endo- or exo-thermal heat energy shifts. Blackbody storage would raise the temperature of our device. Similarly, kinetic energy storage would be obvious as it approached thermonuclear conditions, and melts our electrodes. So, with the one exception of chemical equilibrium shifts where heat storage effect is ambiguous, all of the apparent energy storage modes, produce heat of kinetic motion, which if it approached thermonuclear conditions, would result in something that we do not observe, a melt down.

So there must be some other way of achieving thermonuclear energy storage in a room temperature device. Let's begin our inquiry by looking more carefully at those ambiguous chemical equilibrium shifts. Clearly, the exon-thermal reactions are not a favorable way to store energy. But the endow-thermal process holds some promise.

Heat energy exists in two forms, each of which is described by its own equilibrium modeling framework. The first is a theory of molecular motion in a gas, which bears the revered names of Maxwell and Boltzmann. It relates the heat energy of molecular motion to the system's thermodynamic temperature. Then, in an entirely independent theory, Planck relates the energy of radiant heat, again, to the temperature. However, there is no connection between these two modeling frameworks even though we know that they continually exchange heat, always moving a thermodynamic system to mutual temperature equilibrium, and a lowest entropy state.

When chemical equilibrium shifts occur, a system's chemical constituents and its free energy (radiation domain heat energy) become redistributed into a new lowest entropy state in accordance with the Second Law, and in all irreversible thermodynamic displacements, some of the systems free energy is lost to heat of motion. That lost heat resides in the more primitive of the two heat forms, and it can never be reintroduced so as to reconstruct the previous energy/entropy state.

Nevertheless, it is possible to reverse the direction of an irreversible process by introducing some sort of thermodynamic work. But, when the work is stopped, the process will reverse direction again, and return to its lowest entropy state.

Ilya Prigogine (7) studied these out-of-equilibrium processes, and coined the term 'dissipative structure' to describe a displacement of state, to a non-equilibrium condition, from which there is no return to the previous state. Such a process decreases the system's entropy, and might serve as a basis for models of neg-entropic or far-from-equilibrium transitions similar to those that we see in F&P experiments. In fact, there is already a precedent in cold fusion science for dissipative structures. Hydrogen uptake by a metal hydride lattice is a reversible thermodynamic process exhibiting the dissipative structure's characteristics. Hydrogen nuclei are absorbed into the metal lattice structure, and stored there in a *stable*, far-from-equilibrium manner. Let's see how this might further our understanding of the cold fusion process.

So what is it that locks hydrogen nuclei in the metal lattice, preventing their return to the aqueous phase? The hydrogen nucleus possessed kinetic energy that is no longer apparent once it is absorbed into the metal lattice. Let's for argument's sake, assume that the metal lattice is so large that it has no measurable kinetic energy relative to the tiny proton or deuteron. Let's also assume, that in accordance with the First Law, the proton's initial kinetic energy is completely transferred to the metal lattice. Remember, this is heat energy of molecular motion that is being transferred. Finally, we will assume that the coordinated electrons in the lattice's metal bonds are able to absorb this energy and store it in accordance with the QED definition of a covalent bond; as electromagnetic energy that is continually exchanged between covalent electrons. The reversible process framework allows this kind of transference to a different type of energy. At the same time, each new deuteron that is 'quieted', adds to the radiation domain energy stored as covalent bond energy, thereby moving the radiation domain energy storage one more step away from equilibrium. Later in the loading process, resonant sharing of gamma photons between covalent nuclei facilitates energy storage (12).

Do you see the working of a dissipative structure in this process? The initial kinetic energy is absorbed into the covalent bond structure of the metal lattice where it is stored as radiation domain energy between electrons. This traps it in a stable far-from-equilibrium state, a chemical bond. Furthermore, this is an endo-thermol process, absorbing heat energy of molecular motion from the electrodes surroundings. At the same time, the proton is prevented from moving back into the aqueous phase because, in doing so, it would have to re-acquire its temperature dependent thermal energy, which we might generalize as $k_B T$.

## C. A New, Non-equilibrium Theory of Heat

I have proposed elsewhere (8) a non-equilibrium theory of heat that allows us to follow heat exchange between the radiation and molecular motion domains. But more important for our purposes here, it allows us to quantify their far-from-equilibrium separation. This theory contains two temperatures; one for the mass domain ($T_m$) and another for the radiation domain ($T_R$). These can be separated in a very far-from-equilibrium way as shown in Figure 1. For example, mass domain heating alone, by friction for example, causes the black body spectral curve to move vertically, while radiation domain heating alone moves the spectral curve to the right, but at the same thermodynamic (mass domain) temperature. A normal irreversible heat process causes the curve to move in both directions. A discussion of this theory is contained in (8).

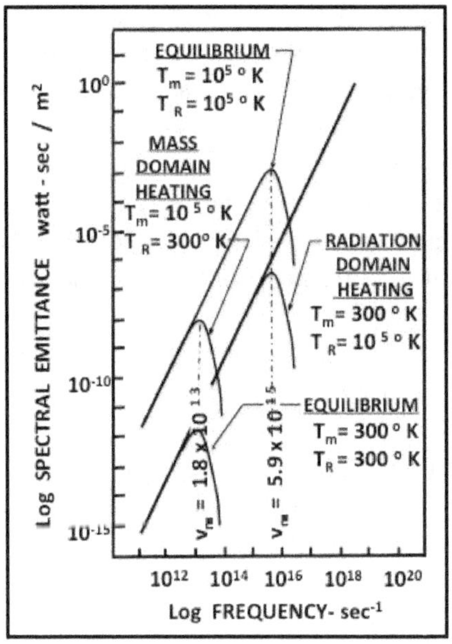


So there are three types of heating that can take place in nature: power accumulation in the mass and radiation domains simultaneously, the normal case, and two far-from-equilibrium heat storage cases. We are now poised to return to the cold fusion question that brought us to this point: How can the cold fusion process accumulate thermonuclear energies in a laboratory temperature device?

I would like to propose that the answer can be found by contrasting the efficiencies of these modes of heating, but with particular reference to the efficiency of radiation domain heating alone. This is most easily seen in Figure 1, which shows the log-log spectral distribution of power emittance for two thermal equilibrium conditions: $300^oK$ and $10^5\ ^oK$, and two very far-from-equilibrium conditions. I want to draw your attention to the two curves at the right hand side of the figure where an equilibrium blackbody at $10^5\ ^oK$ is compared to the non-equilibrium case where the thermodynamic temperature remains at a laboratory temperature of $300^oK$, while the radiation temperature rises to $10^5\ ^oK$. In both cases the effective temperature is $10^5\ ^oK$, but the upper curve's equilibrium representation requires a total power (area beneath the curves) more than two orders of magnitude greater that the far-from-equilibrium condition. In other words, more than 99% of the irreversible process's energy storage is tied up in molecular motion. And yet, the effective free energy available in both cases, the radiation domain energy, is the same. Do you see how radiation domain energy storage is an incredibly efficient way to store free energy?

But there is a more important point to be made here that will be the subject of this paper's next section. In particular, this condition where energy is stored exclusively as radiation domain power, describes the operating conditions for processes that take place at the very limit of the Second Law, where entropy production ceases and the operative thermodynamics are exclusively those of reversible processes.

This is the place where a chemical process in state A, can proceed to state B, but with no loss of process energy to heat of motion. In fact *kinetic motion is foreign to the reversible thermodynamic processes*, unless, and this is important, it occurs in amounts that are so small that they do not impact the reversible process state in any way. We will revisit this constraint later, when we consider the Third Law of Thermodynamics.

If we now continue along this same line of thought, we find that the reversible process domain is the one place in physical theory where the reservoirs of radiant heat energy and heat of molecular motion have no interaction. Kinetic movement is always an entropic process, and therefore foreign to the reversible process domain. This means that the reversible chemical process describing the transition from State A to State B proceeds without any loss of energy to heat of motion, and at least in theory, the chemical equilibrium can shift from state B, back to State A, again, without energy loss or entropy increase, and, this forward and backward chemical shift can recur indefinitely. Do you see the fundamentals of a covalent chemical bond in this description?

This is the domain of the reversible thermodynamic process. It is a 'never land' that we are taught does not exist, and yet it is the only path out of the cold fusion theory dilemma. Furthermore, there is nothing in the laws of physics that precludes this adventure. Indeed, if you carefully critique this presentation, there is no place where it conflicts with experimental evidence or the laws of physics. It is merely an extension of our understanding beyond the current paradigms.

### D. The Reversible Thermodynamic Process in Nature

Let's begin at the beginning. Metal hydrides load at standard conditions in what is known to be a reversible thermodynamic process. These are processes that operate at the very limit of the Second Law of thermodynamics, where there is no energy loss, and the process produces no entropy increase. Another way of viewing the reversible thermodynamic process is that as a result of its operation, there remains no outstanding change in nature.

In fact, it is not possible to view the fundamentals of a reversible thermodynamic process because there is quite simply nothing to look at. There is no energy or matter coming off the process, not even light. These 'visuals' or 'detectables' would be losses from the process, and by definition it would no longer be a reversible process. Thus, the reversible thermodynamic process is essentially hidden from our observation by its very fundamentals.

These processes are generally understood to be only an idealized curiosity in physics that does not exist in the real world. But if we think about it, our experience can recall several common processes that appear to meet our criteria for thermodynamically reversible.

For instance, an atom exhibits all of the reversible thermodynamic characteristics. Its nucleus and electron configuration can remain intact without energy losses or entropy increase for billions of years. The work that the atomic process accomplishes is the stable atomic form. There is one fundamentally important observation that I want you to be clear about in this and all other reversible processes that we will be referencing. It is the absence of any kinetic energy or mass domain events in the reversible thermodynamic process itself. So while the atom may exhibit kinetic behavior, the reversible process that holds it together lies entirely in the radiation domain without recourse to any form of internal thermal motion.

A second example is found in the chemical bond of two or more atoms in a stable molecular form, for example, $\boldsymbol{H_2}$. The covalent bond structure of the molecule is a reversible thermodynamic process that can remain fixed for billions of years with no energy loss or entropy increase. There may be thermal motion of the atomic constituents relative to one another, or kinetic movement of the molecule as a whole, but the internal mechanism, the operative reversible thermodynamic process, exists entirely within the radiation domain. For our purposes, we will view the dynamics of the chemical bond in accordance with QED as the continual exchange of a single photon between two electrons, without energy loss or entropy increase.

Now, I know that this definition of a chemical bond is not exactly that of QED, quantum mechanics, density function models, or empirical models of how chemical bonding works. But, this way of looking at new physics: that it should conform to only what is known (the current paradigm), is counter productive, particularly in cold fusion theory development. We have simply run out of traditional theoretical frameworks to address the anomalous physics seen in our experiments. If we are to advance our understanding of physics, we need to instead ask: Is this new or different concept inconsistent with the physics and experimental results that we are aware of? If the answer is no, some sort of tentative or provisional acceptance should be possible, at least until we see where this nuance takes our understanding of physical principles.

A third example is a photon traveling through the vacuum of space. It exhibits no energy loss or entropy increase. And although it may possess kinetic energy, the photon itself conserves its internal electrical and magnetic energies in accordance with Maxwell's equations, and therefore without internal kinetic energy.

Our fourth example is the reversible process known as Mössbauer resonance (9). Mössbauer found that gamma photons could under certain conditions be absorbed by atomic nuclei that are bound in a metal lattice, and then be re-emitted without loss of energy or entropy increase. The absorption and emission processes were thus, thermodynamically reversible. Similar behavior had previously been observed for x-rays in gasses. Now, I understand that Mössbauer never considered the dynamic state wherein a single gamma photon is continually exchanged between two nuclei. But this behavior is not inconsistent with known physical principles, and lends itself to an understanding of energy storage that will prove useful to our hypothesis of energy accumulation and storage in the cold fusion process.

The last example that I want to bring to your attention was referenced earlier: the loading of a metal hydride lattice with hydrogen or deuterium. It occurs without energy change or entropy increase. But I want you to see how this process is different than the other reversible thermodynamic processes cited above. It includes a kinetic energy component, the deuteron motion, which is followed by a step that is more like the reversible processes above: the transformation of that kinetic energy to radiation domain energy. So let's take a more careful look at the partitioning of the deuteron absorption process.

## E. The Thermodynamics of Energy Accumulation

So what energy does the deuteron possess? There are two kinds. It certainly has kinetic energy, $1/2\mathbf{mv}^2$, and it also has its mass/energy equivelant, $\boldsymbol{E = mc^2}$. At laboratory temperature, only the first of these is accessible. We can then contrast the deuteron's kinetic energy with that of the massive metal hydride lattice where there is virtually no kinetic energy present. At the instant when absorption takes place, the kinetic energy of the deuteron becomes identically zero, but as a consequence of the First Law, the energy is not lost. It is absorbed into the metal lattice structure. And because this process is thermodynamically reversible, the second law requires that the entropy change in the deuteron/lattice complex be zero.

We will begin by recognizing that we are dealing with two thermodynamic systems, not one. One of these exists entirely within the mass domain, where we will be looking at thermodynamic variations resulting from a change in free energy. The free energy/entropy relationship in the mass domain is given by:

$$\boldsymbol{F_m = E_n - T_m S_m} \quad (1a)$$

$$\boldsymbol{dF_m = dE_m - d(T_m S_m)} \quad (1b)$$

$$\boldsymbol{dF_m - dE_m = -T_m dS_m - S_m dT_m} \quad (1c)$$

And it is characterized by; $T_m$ the thermodynamic temperature,
$F_m$ its Helmholtz free energy,
$S_m$ its entropy, and
$E_m$ its total energy.

We use the Helmholtz free energy because the process that we hope to analyze is thermodynamically reversible, and the Gibbs free energy form is inappropriate to that case.

A second thermodynamic system is wholly contained within what I have referred to as the radiation domain. It is characterized by the same four quantities.

(2) $$dF_R - dE_R = -T_R dS_R - S_R dT_R$$

The deuteron absorption process captures the deuteron's kinetic energy from within the mass domain, and absorbs it into the metal hydride lattice where it will be stored as radiation domain energy, and more specifically we will assume that it is stored as electromagnetic bond energy (10).

There are only six unknown differentials in this system of equations. This is so, because the change in both Helmholtz free energy quantities is identically zero. Our process is thermodynamically reversible.

Deuteron absorption into the metal lattice results in a decrease in the mass system's total energy by $\frac{1}{2}mv^2$, the deuteron's kinetic energy, while the radiation domain acquires that energy equivalent.

Moving on to the temperature changes in the mass and radiation domains, we recall that the mass domain's temperature is defined entirely by the kinetic energy of the mass system. So when the deuteron's kinetic energy goes to zero there is an infinitesimal decrease in the mass domain temperature. At that same instant in time, the radiation domain's temperature increases because a single quantum is added to the electrodes internal spectral emittance. We can now calculate the entropy changes.

The molecular motion that specifies the mass domain's temperature is more accurately described as the rate of kinetic energy exchange between molecular quantities. And because temperature measures a derivative, it can be viewed as the rate at which kinetic energy is increased and simultaneously decreased within the F&P cell's deuterium phase. At equilibrium, the total amount of kinetic energy being lost to collisions equals the amount being gained in those same collisions, and there exists equilibrium between the mass and radiation domain's energy content.

This equilibrium is upset by deuteron absorption into the metal hydride lattice. In particular, as the deuteron is 'quieted', there is an infinitesimal decrease in the rate at which kinetic energy is being restored to the mass domain, and the deuteron phase takes a negative kinetic energy step. Some of the mass domain's heat has been lost, $Q_m$ is negative, and the entropy of the mass domain increases;

(3) $$-\frac{Q_m}{T_m} = +S_m$$

When this same energy appears in the radiation domain as a positive change in the rate of radiation emittance and absorption, the radiation domain entropy decreases by that same infinitesimal amount. Nota bene: the non-differential temperature and entropy quantities in both the mass and radiation domains are always positive numbers.

| | Mass Domain | | Radiation Domain |
|---|---|---|---|
| | $dF_m = +dE_m - T_m dS_m - S_m dT_m$ | = | $dF_R = +dE_R - T_R dS_R - S_R dT_R$ |
| Known | + (-) - (+) ? - (+) (-) | | + (+) - (+) ? - (+) (+) |
| | - + | | + - |
| Calculated | - (+)(0 or +) | | - (+)(0 or -) |
| | (0 or +) | | (0 or -) |
| Net | | 0 | |
| | $dF_m = +dE_m - T_m dS_m - S_m dT_m$ | = 0 = | $dF_R = +dE_R - T_R dS_R - S_R dT_R$ |
| Final Tally | + (-) - (+) (+) - (+) (-) | | + (+) - (+) ( -) - (+) (+) |

So, the radiation domain is slightly more organized, taking one small step away from equilibrium. Do you see how the net effect of deuterium absorption into the metal hydride lattice, is in essence, harvesting energy and neg-entropy (10) from the domain of random molecular motion where the entropy increases, and converting it to an entropy decrease, and radiation domain storage of useful free energy that will ultimately be used for thermo-nuclear transmutations.

In reference (10) I go through the calculations that show the deuteron's initial kinetic energy to be about $7.\times 10^{-22}\ ergs$, and that absorption of enough deuterons to fill the surface sites on Fleischmann and Ponn's original experimental electrode would require about $10^{15}$ deuterons, or enough energy transferred to the electrode (1.35MeV) for ignition of deuterium-deuterium fusion. In effect the reversible process responsible for electrode loading (say 30 days) has harvested heat of molecular motion from about a billion ($3.8\times 10^{8}$) deuterons per second to achieve the fusion ignition temperature.

I want you to note that this might explain the very long loading times in Fleischmann-Pons experiments. The ignition requirement is satisfied in $10^{15}$ deuteron absorption steps having extremely small energy increments (in the radio frequency range), and infinitesmal but discrete temporal increments (reversible process steps are sequential).

## F. The Third Law's Relevance

Our quest has now brought us to the Third Law of Thermodynamics: *The entropy change associated with any condensed system undergoing a reversible isothermal process approaches zero as the temperature at which it is performed approaches $0\ ^oK$*. This Law appears to be telling us that the thermodynamic temperature, $T_m$, must go to zero degrees Kelvin for a reversible thermodynamic process to occur. But this is not true. The Third Law acually provides a reference point for the determination of entropy at any other temperature. "The entropy of a system, determined relative to this zero point, is then the *absolute entropy* of that system" (11).

(4) $$S - S_0 = k_b \ln \Omega = k_b \ln 1 = 0$$

Where: $S$ and $S_0$ are the system entropy and the *reference* entropy respectively,
$k_b$ is Boltzmann's constant, and
$\Omega$ is the number of accessible microstates that the process can advance to.

Do you see how it is not necessary that the temperature, and with it the entropy go to zero, but rather, it is a sufficient condition that the entropy *difference*, relative to the reference entropy, goes to zero. This is, by definition, a reversible thermodynamic process. "Hence the initial entropy, $S_0$, can be any selected value so long as all other calculations (of the action) include that as the initial entropy."(11).

I want to now complete our earlier discussion of the Second Law with one final point that becomes relevant here. In *all* reversible thermodynamic processes, every next step is exactly and unambiguously determined by the Principle of Least Action. Thus, the number of accessible microstates, $\Omega$, at any time is identically 1, and the identity in equation (4) is always true for the reversible process. Thus, we say that the reversible thermodynamic process is absolutely deterministic, with no reference to the kinds of probabilistic outcomes that characterize the Second Law's irreversible thermodynamic processes.

Now, let's go back to Figure 1 to see what the Third Law is telling us about the cold fusion process. We will recall that heat is made up of two different processes each existing within its own thermodynamic domain.

The first process is that occurring in the material world, or the mass domain. It contains a certain amount of kinetic energy by virtue of the sensible temperature of the cold fusion device, $T_m$. This is the thermodynamic temperature. It defines the reference state, $S_0$, the *absolute entropy* state of the cold fusion electrode as that prior to *every next step* in the reversible process. It is the entropy of the mass domain system, $S_m$, that must remain constant, insuring that the difference, $S_m - S_0$, remains null during that step. But, because there is no increase in the system's entropy, $S_m$, there can be *no decrease* in the thermodynamic temperature, $T_m$ of the overall cold fusion process. And this conflicts with the thermodynamic analysis in Section E above, where at the micro-scale; the thermodynamic temperature decreased an infinitesimal amount. However, if we are clever we will see the way out of this conflict.

What does $\boldsymbol{S_m - S_0 = 0}$ really mean? The earlier thermodynamic analysis indicated an infinitesimal decrease in the mass domain's temperature as a deuteron is absorbed. But, we now find that there can be no decrease in $T_m$. So, if $S_m - S_0$ is to remain null, there appears to be an allowable departure from the mass domain's *absolute entropy* state, a very small entropy quantity:

(5) $$-\frac{\Delta Q_m}{T_m} = -\frac{(-\frac{1}{2}mv^2)}{T_m} = (+)\Delta \boldsymbol{S_m}$$

that must be restored continually to satisfy the reversible process constraint that $(S_m - S_0)$ remains *essentially* zero. I don't know what this limit is, but I do know that it sets the maximum energy increment that can be absorbed in a single deuteron absorption step, thereby limiting energy harvesting to small kinetic energy increments that are, at least according to this theory, in the RF range. It is, I believe this small increment that will help define operating conditions for commercial cold fusion devices.

Furthermore, this observation also shows how the cold fusion process fails the test for a perpetual motion machine. In particular, the decrement in kinetic energy must be restored after every process step. This heat can be derived from the electrical current in the electrode, and even by recycling heat from the nuclear process. Its speed is the speed of light.

I have identified a second place where a similarly small energy quantity needs to be computed. In particular, there is a limit to the velocity at which hydrogen nuclei can move from their absorption site into deeper lattice locations. This energy too, needs to be below a permissible threshold. I don't know what this threshold is, or if it might impact design criteria for a cold fusion prototype.

Finally, regarding the second thermodynamic system, denoted by the temperature, $T_R$. The radiation and mass domains normally exchange heat energy, always seeking an equilibrium state where $T_R = T_m$. However, because the mass and radiation thermodynamic systems are separate within the context of the reversible thermodynamic process, it becomes possible to increase the radiation domain's free energy storage while holding the mass domains energy constant. In this way the two can be separated in both their energy content, and their entropy state in a very far-from-equilibrium way. More importantly for our purposes, if we store the radiation domain energy as chemical bonds, it becomes possible to make this separation stable even as the two thermodynamic systems move to a very far-from-equilibrium separation. Thermonuclear energies and temperatures become possible when photon energy storage between covalent electrons makes a transition to resonant, gamma photon, storage between Mössbauer resonant nuclei, and ultimately, the ignition energy for nuclear transmutations. This far-from-equilibrium state was the subject of my paper at ICCF-19 (12), where I discuss the atom's temperature, and its consequences to a theory for overcoming the coulomb barrier.

## G. References


[1] Hagelstein, P. L., "Bird's Eye View of Phonon Models for Excess Heat in the Fleischmann-Pons Experiment", J. Condensed Matter Nucl. Sci. 6, (2012) 169-180.

[2] Widom, A, Larsen, L, "Ultra Low Momentum Neutron Catalyzed Nuclear Reactions on Metallic Hydride Surfaces", European Physical Journal C – Particles and Fields, Vol. 46(1) (2006), 107-110.

[3] Muelenberg, A., Sinha, K.P., "Deep-electron Orbits in Cold Fusion", J. Condensed Matter Nucl. Sci. 13 (2014) 368–377

[4] Szumski, D. S., "Rethinking Cold Fusion Physics", Infinite Energy, 20, 120, 2015.

[5] Szumski, D. S., "Cold Fusion and the First Law of Thermodynamics", Infinite Energy, 20,122, 2015.

[6] Szumski, D. S., "Can We Explain Excess Heat Uncertainty with a Law of Physics – An Essay", Accepted by Infinite Energy, 21,128, 2015. (available at ww.leastactionnuclearprocess.com)

[7] Prigogine, I., Self-Organization in Non-Equilibrium Systems, Wiley, 1977.

[8] Szumski, D.S., "Consequences of Partitioning the Photon into its Electrical and Magnetic Vectors upon Absorption by an Electron", In the Nature of Light: What are Photons? V. Chandrasekhar Roychoudhuri; Al F. Kracklauer; Hans De Raedt, editors, Proceedings of SPIE Vol 8832 (SPIE, Bellingham, WA), 883201, 2013.

[9] Mössbauer, R. L., Recoilless Nuclear Resonance Absorption of Gamma Radiation", Nobel Lecture, December 11, 1961.

[10] Szumski, D.S., "Nickel Transmutation and Excess Heat Model Using Reversible Thermodynamics", J. Condensed Matter Nucl. Sci. 13, 554–564, 2014.

[11] 'Third law of thermodynamics' in Wikipedia, Retrieved 11/15 from http://en.wikipedia.org.

[12] Szumski, D.S., "The Atom's Temperature", submitted to J. Condensed Matter Nucl. Sci.

## Chapter 9

# Design of a High Performance Cold Fusion Electrode

### A. Introduction

The cold fusion dream seems close at hand. But, a careful look at the current landscape reveals broad areas of uncertainty, where the only navigable path is trial-and-error experimentation. And what makes this observation all the more striking is the realization that there has been very little change in the theoretical portion of that landscape over the last 27 years. During the past 10 years, nearly all of the scientific advances have been in the experimental domain; and much of that work has focused on electrode design.

The Fleischmann-Pons effect is related more to the electrode composition and its manufacture than any other factor. For years we had seen the importance of electrode materials in the way there were preferred suppliers, whose wires and foils were known to produce positive excess heat results. More recently, electrodes are designed to eliminate long loading times (1), to explore experimental curiosities and advance theoretical understanding (2), or to develop ideas aimed at enhancing excess heat gains(13). But the one obstacle to effective electrode design continues to be our failure to understand cold fusion at the level of its physics fundamentals.

There are reasons for this. First, cold fusion behaves like no other physical process that we have ever encountered. Its results are not reproducible. In the most extreme case there is randomness to the presence or absence of excess heat. This behavior is unprecedented in physics. However, beyond this there are other unexplained process characteristics that confound reproducibility:

1. The non-reproducible, and seemingly random time history of excess heat,
2. Heat's cessation and renewal during a single experiment,
3. Heat evolution even after electrolysis power is turned off, and
4. The inevitable, but unpredictable termination of the excess heat response.

The need for a cohesive theory becomes more important as engineers attempt to develop commercial prototypes. It is in the absence of theoretical understanding that development has to default to trial and error experimentation in search of a magic formula or an accidental breakthrough.

This paper addresses some of the fundamental issues that will be involved in commercial electrode design. It will then use the Least Action Nuclear Process [LANP] theory of cold fusion to illustrate an electrode design methodology.

Toward this end, the paper has four parts. In the first, we will show how using excess heat as the primary experimental variable is of little or no benefit in understanding the fundamental physics of cold fusion. In the second, we will look at the source of the seemingly random variability in our experiments, and endeavor to find its source. Third, it will be necessary to identify deficiencies in the current data set, and develop an experimental program to collect the data required to calibrate a cold fusion model, and prepare it for electrode design. The fourth and final part of this paper will illustrate how the LANP theory can be used as an electrode design tool to optimize commercial heat production, and even to design an electrode with Mars Mission reliability.

## B. The Relationship Among Cold Fusion Variables

The emphasis on measuring excess heat as the primary variable in experimental work is not a very productive avenue for understanding the cold fusion process. Heat is a secondary, or possibly tertiary effect. It is merely the consequence of the more fundamental process: nuclear transmutation (3,4). If we are to understand the cold fusion process at its fundamental level, we need to keep our attention and our measurements focused on those nuclear transmutations; and more specifically their nuclear chemistry, and their temporal sequence.

To understand why using excess heat is so limiting as a primary experimental variable, we have to look at it in its relationship to other cold fusion variables. These are summarized in Table 1.

Table 1 – Three Levels of Cause/Effect Variables in the LANP Cold Fusion Process

| |
|---|
| Primary independent variables |
| Energy storage (in chemical and Mossbauer bonds) |
| Temperature (thermonuclear is ultimately required) |
| Secondary variables |
| Nuclear transmutations |
| Tertiary variable |
| Excess heat |

When we look at cold fusion through this lens we see that excess heat is not among the process fundamentals. It is a product of nuclear transmutations that result from thermonuclear energy and temperature conditions (10). And although these thermonuclear conditions are not apparent in our experiments, we know that they exist there because we observe the resulting nuclear transmutations (2,8,9). The First Law requires this, and there is no process in the natural world that will ever violate this principle.

The only way out of this dilemma is the Least Action Nuclear Process [LANP] theory of cold fusion. It is the only theoretical framework that proposes mechanisms by which thermonuclear conditions can occur in a laboratory temperature experimental device (5). We will be using this theory to illustrate the source of variability in cold fusion experiments, and then using this understanding to show how a reliable cold fusion electrode might be designed.

## C. The Source of Experimental Variability

It is informative to first look at the *types of variability* in each of the first two stages of the cold fusion process, to understand variability in the third. Because we are using the LANP theory, we are assuming that the cold fusion process occurs in accordance with the rules of reversible thermodynamics.

1. Variability in Energy Storage and Internal Process Temperature

According to LANP theory, energy uptake and storage occurs in discrete, microscopic energy steps (4), and these produce correspondingly small incremental temperature changes (5). Both the energy storage mechanism, and the process temperature are internal to the device, and therefore, hidden from our observation (5). This occurs because there is no change in nature resulting from the operation of a reversible thermodynamic process. There is quite simply nothing to see or measure because no mass or energy change occurs as a result of the reversible processes operation.

LANP process energy begins as kinetic energy of deuterium molecules that is harvested from the environment, and transformed to what I have called radiation domain energy (4). But, there can be no kinetic energy within the reversible thermodynamic process environment (6). Kinetic energy introduces statistical uncertainty, something that is foreign to reversible thermodynamics. Thus, the only way that the hypothesized reversible thermodynamic process can occur in an irreversible world is if there is a maximum allowable kinetic energy increment at each step in the process; a small numerical limit, below which, kinetic energy can be present in an imperfect reversible process without affecting its continuance (6). I believe that this energy limit is temperature dependent, on the order of $k_b T$, but of lower magnitude. An analysis of the cold fusion process relative to the Third Law (6) describes the rationale for this numerical limit.

These two independent variables: energy storage and process temperature lie entirely within the reversible process domain where all events are discrete and completely deterministic (6). Consequently, they contribute no variability to excess heat measurements.

2. Nuclear Transmutation Variability

Nuclear transmutations are similarly precise, and strictly deterministic. LANP theory finds that the transmutation sequence is stepwise, and that every 'next step' in the reversible process is precisely and unambiguously determined by the Principle of Least Action (4). So again, this is a strictly deterministic variable. However, *and this is what is important here*, that 'next step' in the reversible process represents the smallest mass/energy change, *regardless of its sign*. And because the + or - direction of the energy change is itself random, a high degree of variability is introduced into what is otherwise a strictly deterministic process (7). And, as it turns out, these sign changes, are the sole source of randomness in the cold fusion process's excess heat time series.

In this way, we can isolate *all* of the variability in the cold fusion process to the random direction of discrete + and – mass/energy changes. Aside from this one source of variability, cold fusion fundamentals are exactly deterministic, and indeed, even the direction of mass/energy change is completely deterministic. Thus, the cold fusion process operates precisely the way that laws of physics are supposed to behave; strict determinism. In this way the LANP theory addresses the most convincing criticism (11) that mainstream physics uses to discredit cold fusion physics.

## D. Expanding the Current Data Set

The current data set is sufficient to support the nuclear origins of excess heat. Mizuno (9), Miley (8), and Iwamura (2) have documented nuclear transmutations beyond any reasonable doubt. An analysis of these data (4) found that Miley's data represent an intermediate condition in the evolution of nuclear transmutations, and differs from Mizuno's results in one important way. Mizuno's data are from longer term experiments, wherein the final electrode's composition contains virtually every stable isotope. Miley's data contains only a subset of the total stable isotope catalog. Together, these two data sets show how transmutations do not occur across the board, but rather according to some temporal order.

We need to expand this limited nuclear data set to, among other objectives, learn more about the temporal sequence of nuclear transmutations. McKubre has noted the absence of this kind of rigorous tracking experiment (10).

The purpose of the laboratory investigation that is about to be proposed is two-fold. First it will fill a serious deficiency in the current data set by collecting systematic nuclear transmutation time series data that lends insight into the fundamentals of the cold fusion process. It is hard to imagine that this field of study has advanced so far without a comprehensive data set on such an important variable: nuclear transmutations.

The LANP model requires excess heat measurements and nuclear transmutation time series data for its calibration. Nevertheless, while conducting a laboratory experiment on this scale, it only makes sense that other perceived data needs be collected so that alternative theories can be refined and tested. This data set will, furthermore, establish a baseline condition against which all past and future data can be referenced.

Such an experiment should run five identical experiments from a common power source, using nickel as the electrode material. The NANOR might even be well suited to this experimental protocol. Measurements might include:

1. Excess heat calorimetry
2. Electrode isotope composition: one before, three or four during, and one after excess heat cessation
3. Cumulative gas phase analysis at each of five sampling times
4. Radiation signature measurements: RF, microwaves, far/near IR, UV, X-ray, and Gamma
5. Neutron measurements
6. AA analysis to confirm SIMS measurements
7. Other measurements

A second experiment using palladium as the electrode material might follow this initial experiment.

While it would be convenient that the experiment produces excess heat, this may not be required (7). According to the LANP theory, the measurements that we are interested in are operative regardless of the excess heat condition. Mizuno has collected transmutation data for an electrode that produced no excess heat (12).

Nickel is being proposed as the electrode material for two reasons. First, many of the required energy change calculations have already been completed for nickel isotopes (4). And, because the required calculations are electrode specific, the impurities in this study's electrodes will have to be partially recalculated. Secondly nickel is the most likely candidate for commercial electrodes. Without computerized calculations, it takes hundreds of hours to calculate mass/energy changes, and expected stable isotopes for a single electrode composition. This is a process that needs to be automated.

The LANP model is developed to the point where it should be able to make rough predictions (without an LANP computer program), or accurate predictions (with an LANP computer program), for the order of new isotope appearance. In this way, this experiment will serve as a calibration of the LANP model, and a test of some of its predictions including:

1. The nuclear nature of the process,
2. A rough, or precise, order of isotope appearance,
3. 'No heat' nuclear transmutations (if no excess heat is observed),
4. Possibly a very rough cut, (or accurate prediction) on heat/no-heat dynamics
5. Differences at the final sampling will show that additional nuclear transmutations occur even in the absence of excess heat.
6. Surface hardness changes in pre- and post-experiment electrodes should occur.

## E. Electrode Design

$$ {}^{61}_{28}Ni + {}^{107}_{47}Ag + (2){}^{2}_{1}H \xrightarrow{fusion} {}^{172}_{70}Yb $$

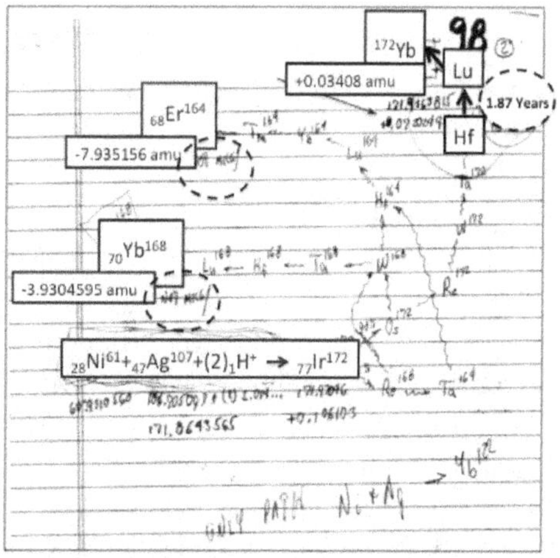


We will begin by reviewing the transmutation selection process in the LANP model. Because transmutations occur within the model's reversible thermodynamic space, their every 'next step' is governed only by the Principle of Least Action, which in this case means: lowest mass/energy change. Consider the nuclear transmutation:

$$ {}^{61}_{28}Ni + {}^{107}_{47}Ag + (2){}^{2}_{1}H \xrightarrow{fusion} {}^{172}_{70}Yb $$

The normal decay sequence for this reaction has four paths, two of which converge. It results in three stable isotope end products, but according to this theory, the only transmutation product is ${}^{172}_{70}Yb$ because it results in the smallest mass/energy change: +0.03408 amu. Also note that this change occurs even though one of the intermediate reaction steps:

$$ {}^{172}_{72}Hf \xrightarrow{Electron\ capture} {}^{172}_{71}Lu\ ,\ t_{1/2} = 1.87Yr $$

has a half life far longer than the experiment. See www.LeastActioNuclearProcess for this and other transmutations paths.

If we array all of the possible nuclear transmutations for a specific electrode and its impurities (4) in this same way, patterns become evident:

1. Both positive and negative heat changes occur,
2. The greater the number of deuterons reacting with the electrode material and/or its impurities, the higher the absolute value of the mass/energy change. The electrode thus produces one and two deuteron reactions first, and may never get to those less probable 5 and 6 deuteron reactions.
3. Based on an inventory of final electrode isotope composition, there is evidence that the following kinds of reactions occur:
   a. Deuteron fusion reactions with base electrode material ($Ni^{58}_{28}, Ni^{60}_{28}, Ni^{61}_{28}, Ni^{62}_{28}, Ni^{64}_{28}$) plus $n[d^2_1]$
   b. Deuteron fusion reactions with electrode impurities ($Ag^{107}_{47}, Ag^{109}_{47}, Cu^{63}_{29}, Cu^{65}_{29}, Zn^{64}_{30}$...) plus $n[d^2_1]$
   c. Fusion of base electrode material and impurities ($Ni^{58}_{28} + Ag^{107}_{47} + n[d^2_1], \ldots$)
   d. Fission reactions, and
   e. Alpha decay reactions that yield $He^4_2$ which is observed in some experiments.
4. Eventually, the electrode succumbs to the great entropy principle, and transmutations slow to the no effect level. The electrode is being depleted.
5. Nuclear transmutations appear to occur without regard to half-life constraints. As long as a transmutation product is stable, and it results in the smallest energy change, it occurs at the 'next step'.
6. At each 'next step' in the process, transmutation products are subject to all of the possibilities listed for the base electrode material and its impurities.

Number 6 is the problem. We might be able to design an electrode that is doped to yield exothermal reactions, at least initially. But, because new transmutation products are continually appearing in the electrode, the Least Action Nuclear Process is continually changing. Sometimes this continues to favors exothermal reactions. Sometimes it does not.

Before we go further, do you see how we can calculate in a very precise manner, the Least Action Nuclear Process for a specific electrode composition? It is exactly deterministic. It depends only on the next lowest energy change in the electrode, and this can be known with the precision and accuracy that we expect in physics. And, that 'next step' also includes the amount of heat that is released(+) or consumed(-) during that step. A computer program needs to be developed to keep the electrode isotope inventory, and the energy changes for all pending nuclear transmutations. These inventories, and the heat content in the cold fusion reactor need to be updated at each 'next step' in the integration. In this way it is possible to predict the excess heat time series with the accuracy that we expect of physics.

Ultimately, the design secret lies in modifying the initial electrode composition, such that the initial composition and its 'later generation' isotopes yield predominantly exothermal reactions (or reactions that favor other design based objectives). This is accomplished in several ways.

1. Element doping
2. Isotope doping
3. Base metal isotope adjustment

However, electrode composition alone will not result in a continuous process that conforms to specified design objectives, necessitating electrode operational interventions including, for example:

1. Continuous (possibly variable) flow of electrode particles
2. Dynamic electrode substitutions
3. Electrode additions/deletions
4. Multiple electrodes to 'smooth out' amplitude effects and output variability
5. Electrode re-plating to allow recycling in resource limiting circumstances

Electrodes can also be designed for purposes other than excess heat production:

1. Production of specific elements or isotopes
2. Radioactive waste stabilization

These methods can be used individually or in suitable combinations to produce precise, and reproducible results. A few examples will illustrate this facility of the Least Action Nuclear Process.

We will begin by recalling that the design calculations are **electrode specific**. Each new isotope in the electrode brings with it an array of possible reactions with both deuterons and other electrode isotopes. And in all cases, the sequence of nuclear transmutations, occur in the order of least energy change. We will call this sequence of nuclear reactions the **Least Action Nuclear Process** *for this specific electrode*. If two or more electrodes have precisely the same initial, homogeneous isotope composition, their transmutation sequence, and their excess heat time series will be identical. If we then make precisely the same change in each of those initial electrode compositions, the Least Action Nuclear Process, its transmutations, and its excess heat outputs will be altered, but identically the same for each electrode. Do you see how the precision and reproducibility found elsewhere in physics, exists within this cold fusion process? And do you see how we can use the LANP model to predict outcomes with that same precision?

This is what we want to do as we design electrodes for a specific purpose, whether it be excess heat production, the synthesis of specific elements or isotopes, or another objective.

So what is the next step? First, we need to develop a computer program for performing the tedious and repetitive calculations required to analyze any specific electrode's isotope composition. I have already done these calculations (4), and consequently I can be of assistance here. Remember, each new electrode requires a new set of candidate nuclear reactions. Then we have to calculate the precise Least Action Nuclear Process path for that electrode, including a prediction of its excess heat profile, or other process objective.

Now that we have the Least Action Nuclear Process programmed, it becomes possible to run the program to arrive at a good initial electrode composition that produces, for example, commercial excess heat ratios. We might then run the LANP program to the point where endo-thermal processes begin to predominate. By inspecting the mass/energy changes occurring at that time, we can introduce into our hypothetical electrode, isotopes that produce their first exothermal LANP reactions at that mass/energy change, thereby extending the period of excess heat production. However, the great entropy principle, the Second Law of Thermodynamics, will eventually bring this adventure to an end, and the spent electrode will have to be replaced or renewed. What we have done, is a set of calculations that predict exactly and unambiguously the outcome of our modifications to the initial electrode...something that we then manufacture exactly.

Renewal might involve electrode replacement, electroplating over a spent electrode with a new recipe for excess heat production, or spitting new isotopes onto its surface to achieve new excess heat avenues. Replacement might be as simple as feeding new electrode material into the reactor as wire, foil, or even a continuous flow through system of metal hydride micro-spheres.

There's more. We have not yet addressed the issue of quality assurance. For example, if your industrial process only allows heating within specific bounds, as might be the case in a spacecraft, you want to design the electrode, using a combination of electrodes that, in total, produce the desired excess heat profile. For example, a second electrode might be run simultaneously to fill in low spots in the excess heat profile. If very tight control on heat variability is required, many electrodes might be run simultaneously to achieve an average condition that meets planning objectives. There might even be supplemental preloaded electrodes that are introduced to the reaction space when needed, or retracted from it, to maintain heat production within specific bounds. The reactor might also be run in feedback mode. Feedback loops might be monitored to insert and retract specific 'control rods' to eliminate unwanted process variations.

Similar design and operating features might be employed to throttle, or reduce excess heat production, or any other LANP objective. For example, electrodes might be designed to favor the production of specific rare earth metals or noble gasses, perhaps even precious metals.

Let me illustrate the LANP electrode design process for a hypothetical electrode. Figure A presents the excess heat profile for a hypothetical electrode with isotope impurities: $^{28}_{14}Si, ^{29}_{14}Si, ^{30}_{14}Si, Ag^{107}_{47}, Ag^{109}_{47}, Cu^{63}_{29},$ ..., etc. The excess heat profile shows a sharp increase in week 1, and two decreases in week 2, There is also a small recurrence of excess heat in weeks 3 and 4. Excess heat is calculated as the heat evolved from exo-thermal nuclear transmutations minus the heat consumed by endo-thermal nuclear transmutations during any time period:

$$Excess\ heat = q_{exothermal} - q_{endothermal}$$

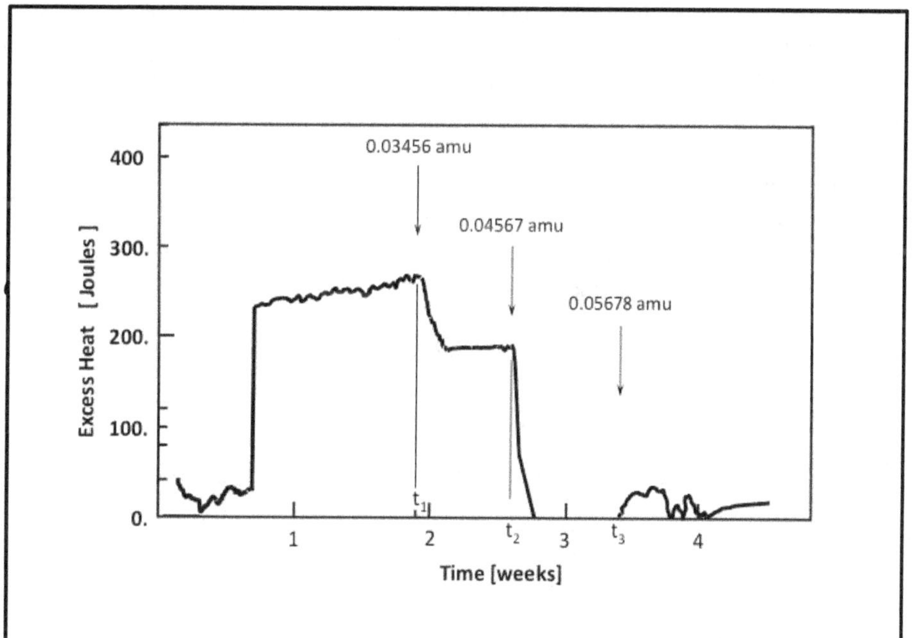


(A)

Exothermal reactions predominate during most of weeks 1 and 2, although at two different levels. Then the excess heat profile plummets as endo-thermal transmutations begin to predominate toward the end of week 2. The design engineer may decide that the first decrease in excess heat is tolerable. He might then use isotope addition **1** to the electrode that sponsors exo-thermal transmutations at a mass/energy change of 0.0456 amu. The result is shown in Figure B where the decline in excess heat is now changed to excess heat production of about 290 Joules. You should also note that the excess heat that had occurred in week three has now been pushed forward in time, and appears as an addition to the excess heat response late in week 3.

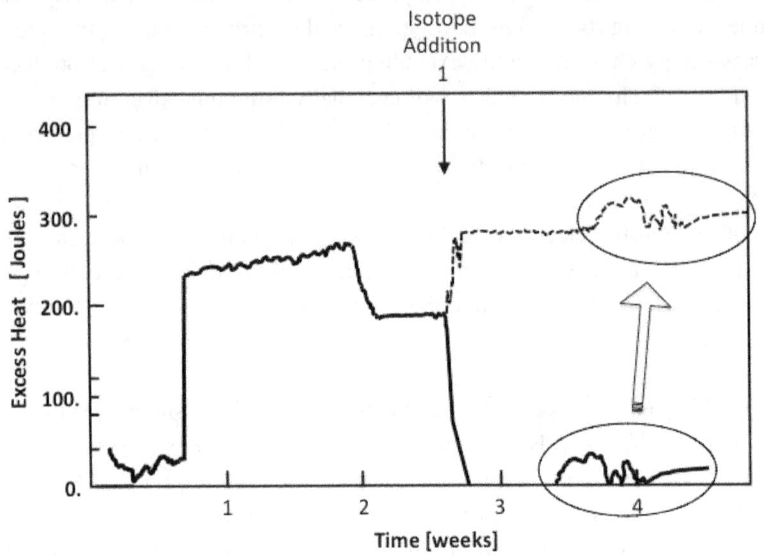


**(B)**

You are probably envisioning the change to the original electrode occurring in a laboratory where a metallurgist is creating a new electrode for experiment B. That is not the way that it is done. We are using our LANP computer program to simulate the original excess heat profile from only an analysis of the original metal electrode's isotope composition. Then we are allowing the program to introduce new electrode isotopes, and calculate the projected excess heat profile.

Look at Figure C where the design engineer has now decided to eliminate the first drop in excess heat that began at the end of week two by adding isotope 2 to the electrode. This trial yields some short term excess heat production, but that too dies off at the end of week 3.

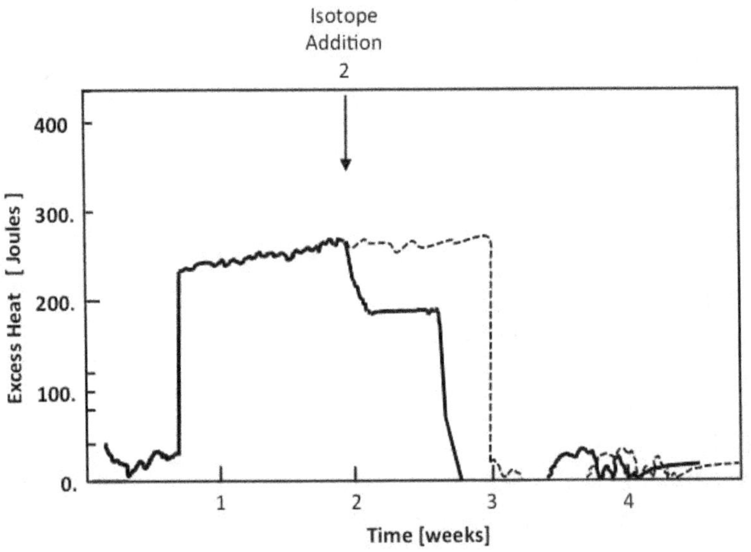


**(C)**

Figure D is the designer's electrode performance following several more simulations with the LANP design program. The design has arrived at a combination of isotope additions that provides a favorable excess heat time series until the end of week 4. There are limits to how long this excess heat profile can persist. The second law of thermodynamics is continually acting on the *imperfect* reversible thermodynamic process to bring about excess heat cessation. Other factors such as electrode deterioration from surface cracking, and fouling with useless electrode transmutation products .

Using this methodology, John O'M. Bockris might someday even find amusement making gold where treasury agents cannot find him (14).

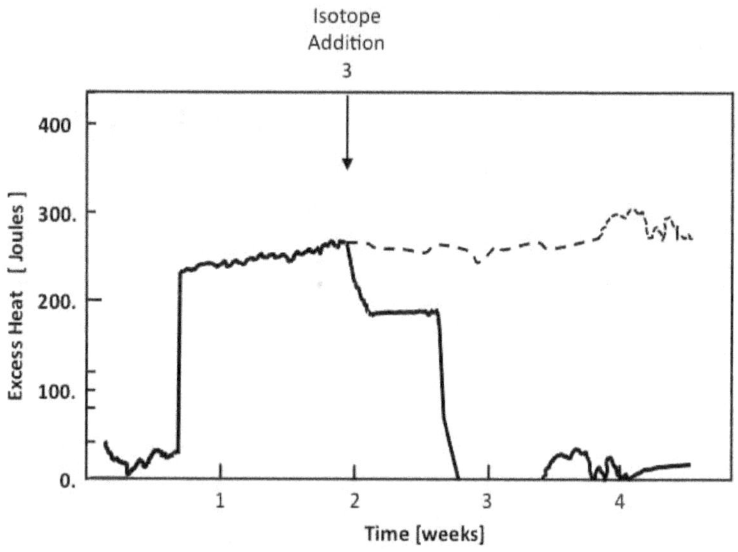


(D)

## F. References


1. Szpak S., Mosier-Boss, P.A., eds., Thermal And Nuclear Aspects Of The Pd/D**2** O System, Thermochimica Acta 410 (2004) 101–107
2. Iwamura, Y.,M. Sakamo, T. Itoh, Elemental Analysis of Pd Complexes: Effects of $D_2$ Gas Permeation, Jpn. J. Appl.Phys. 41, 4642-4648, 2002.
3. Mizuno, T. , Nuclear Transmutations: The Reality of Cold Fusion, US: Cold Fusion Technology, 1998.
4. Szumski, D.S. "Nickel Transmutation and Excess Heat Model Using Reversible Thermodynamics", J. Condensed Matter Nucl. Sci. 13 (2014) 554–564
5. Szumski, D.S., "The Atom's Temperature", presented at the 19th International Conference on Condensed Matter Nuclear Science, 2015, Padua, Italy. available at: www.LeastActionNuclearProcess.com.
6. Szumski, D.S., "Cold Fusion and the Three Laws of Thermodynamics", submitted to Infinite Energy magazine, 1/2/16.
7. Szumski, D.S., "Can we Explain Excess Heat Uncertainty with a Law of Physics, Infinite Energy Magazine, Issue 128, July/August
8. Miley, G., J Patterson, "Nuclear Transmutations in thin-Film Nickel Coatings Undergoing Electrolysis", J. New Energy, vol. 1, no. 3, pp. 5-38, 1996.
9. Mizuno, T., T. Ohmori, and M. Enyo, "Anomalous Isotopic Distribution in Palladium Cathode After Electrolysis", J. New Energy, 1996. 1(2): p. 37.
10. McKubre, M.C.H. and F.L. Tanzella, Cold Fusion, LENR, CMNS, FPE: One Perspective on the State of the Science Based on Measurements Made at SRI. J. Condensed Matter Nucl. Sci., 2011. **4**: p. 32-44.
11. Shanahan, K.L., Comments on "A new look at low-energy nuclear reaction research", J. Environ Monit. 12 1756(2010).
12. Szumski, D.S., personal communication, analysis of T. Mizuno's data
13. Nagel, D., Energy Gains from Lattice-enabled Nuclear Reactions, Current Science, vol. 108, No 4, 2015.
14. Bockris, J.The History of the Discovery of Transmutations at Texas A&M University. In the Eleventh International Conference on Condensed Matter Nuclear Science.2004, Marseille, France.

# Appendix A

Derivation of :

$$f_3\left(v_1/v_m\right) = \left(v_1/v_m\right)^2\left[\frac{1}{2}\left(\ln\left(\frac{v_1}{v_m}\right)\right)^2 - \frac{3}{2}\ \ln\left(\frac{v_1}{v_m}\right) + \frac{7}{4}\right]$$

$$f_n\left(v_1/v_m\right) = \iint \left[\ln\left(v_1/v_m\right)\right]^2 d\left(v_1/v_m\right)$$

Substituting:

$\left(v_1/v_m\right) = x$

first integration:

$$\int (\ln x)^2 dx = x(\ln x)^2 - 2x(\ln x) + 2x$$

$$\therefore f_2 = \quad x(\ln x)^2 - 2x(\ln x) + 2x$$ $2^{nd}$ Integral

second integration:

$$\iint (\ln x)^2 = \int\left[x(\ln x)^2 - 2x(\ln x) + 2x\right] dx$$

$$\int x(\ln x)^2 dx = \frac{x^2}{2}(\ln x)^2 \ - \int x(\ln x)\, dx$$

$$-\frac{1}{2}x^2(\ln x) \quad +\frac{1}{4}x^2$$

$$-\int 2x(\ln x)dx = \quad -\frac{2}{2}x^2(\ln x) \quad +\frac{2}{4}x^2$$

$$+\int 2x\, dx = \quad +\frac{2}{2}x^2$$

---

$$\therefore f_3 = \quad \frac{x^2}{2}(\ln x)^2 \quad -\frac{3}{2}x^2(\ln x) \quad +\frac{7}{4}x^2$$ $3^{rd}$ Integral

# Appendix B

## Analysis of Miley's LANP Data for Nickel Microspheres

[Presented in order of increasing Least Action]

Table 1 - Nuclear Reactions from Miley's Nickel Data – Ordered in LANP Order

| Nuclear Reaction | Initial Isotope | Stable Isotope | Energy Change (amu) |
|---|---|---|---|
| $(2)^{60}_{28}Ni \xrightarrow{fusion}$<br>$^{120}_{51}Te \xrightarrow{fission}$ | $^{120}_{56}Ba$ | $^{116}_{50}Sn, ^{118}_{50}Sn, \left[^{120}_{51}Te\right]$ **absent**<br>$(2)^{60}_{28}Ni$ | +0.0424482<br>**0.0000000** |
| $^{62}_{28}Ni \xrightarrow{fission}$<br>$^{62}_{28}Ni - ^{4}_{2}He \xrightarrow{\alpha}$<br>$^{62}_{28}Ni + ^{2}_{1}H^{+} \xrightarrow{fusion}$<br>$^{62}_{28}Ni + (2)^{2}_{1}H^{+} \xrightarrow{fusion}$<br>$^{62}_{28}Ni$ | $(2)^{31}_{14}Si$ | $(2)^{31}_{15}P$ **absent**<br>$^{58}_{26}Fe$ **absent**<br>$^{64}_{28}Ni$ **absent**<br>$^{66}_{30}Zn$ **absent**<br>$^{62}_{28}Ni$ **(note 6)** | +0.0191782<br>+0.0075337<br>-0.01448087<br>-0.3051524<br>**0.0000000** |
| $(2)^{61}_{28}Ni \xrightarrow{fusion}$<br>$^{122}_{52}Te \xrightarrow{fission}$ | $^{122}_{56}Ba$ | $^{118}_{50}Sn, \left\{^{122}_{52}Te\right\}$ **absent**<br>$(2)^{61}_{28}Ni$ | -0.0409319<br>**0.0000000** |
| $(2)^{60}_{28}Ni + (3)^{2}_{1}H^{+} \xrightarrow{fusion}$ | $^{126}_{59}Pr$ | $^{125}_{52}Te, \left[^{126}_{54}Xe \uparrow\right]$ | **0.0003958** |
| $(2)^{58}_{28}Ni + (3)^{2}_{1}H^{+} \xrightarrow{fusion}$<br>$^{122}_{52}Te \xrightarrow{fission}$<br>$^{122}_{52}Te - ^{4}_{2}He \xrightarrow{\alpha}$ | $^{122}_{59}Pr$ | $^{118}_{50}Sn, ^{120}_{52}Te, ^{121}_{51}Sb, \left[^{122}_{52}Te\right]$ **absent**<br>$(2)^{61}_{28}Ni$<br>$^{118}_{50}Sn$ | -0.0099461<br>-0.050878<br>**+0.001162** |
| $^{62}_{28}Ni + (1)^{2}_{1}H \xrightarrow{fusion}$<br>$^{64}_{30}Zn \xrightarrow{\beta^{+}\beta^{+}}$<br>$^{64}_{30}Zn \xrightarrow{fission}$<br>$(2)^{32}_{15}P \xrightarrow{\beta^{-}}$<br>$(2)^{32}_{16}S \xrightarrow{fission}$ | $^{64}_{29}Cu$<br>$(2)^{32}_{15}P$ | $^{64}_{28}Ni$<br>$^{64}_{30}Zn$ **absent(note 1)**<br>$^{64}_{28}Ni$<br>$(2)^{32}_{16}S$<br>$(4)^{16}_{8}O$ **Example** | -0.014480<br>-0.013304<br>-0.014480<br>+0.005368<br>**+0.001696**<br>+0.037212 |

| | | | |
|---|---|---|---|
| $(2)^{58}_{28}Ni \xrightarrow{fusion}$<br>$^{116}_{50}Sn \xrightarrow{fission}$ | $^{116}_{56}Ba$ | $^{111}_{48}Cd, ^{112}_{50}Sn, ^{114}_{50}Sn, ^{115}_{50}Sn, \left[^{116}_{50}Sn\right]$**abse**<br>**nt**<br>$(2)^{58}_{26}Fe$ | -0.0310552<br>**-0.0020673** |
| $^{109}_{47}Ag - (1)^{4}_{2}He \uparrow \xrightarrow{\alpha}$<br>$^{105}_{46}Pd \xrightarrow{fission} {}^{52}_{23}V$<br>$\xrightarrow{fission} {}^{53}_{23}V$ | $^{105}_{45}Rh$ | $^{105}_{46}Pd$ **(note 2)**<br>$^{52}_{24}Cr$<br>$^{53}_{24}Cr$ | **+0.002936**<br>-1.021133 **H**<br>+0.097915 **H** |
| $^{107}_{47}Ag - (1)^{4}_{2}He \uparrow \xrightarrow{\alpha}$<br>$^{103}_{45}Rh \xrightarrow{fission} {}^{51}_{23}V$<br>$\xrightarrow{fission} {}^{52}_{24}Cr$<br>$^{103}_{45}Rh \xrightarrow{fission} {}^{50}_{23}V$<br>$\xrightarrow{fission} {}^{53}_{24}Cr$ | $^{103}_{45}Rh$<br>**(stab**<br>**le)** | $^{103}_{45}Rh$ **absent(note2)**<br>$^{51}_{23}V$<br>$^{52}_{24}Cr$<br>$^{50}_{23}V$<br>$^{53}_{24}Cr$ | **+0.003039**<br>}-0.018026<br>}-0.017053 |
| $^{64}_{28}Ni - (1)^{4}_{2}He \xrightarrow{\alpha}$ | $^{50}_{26}Fe$ | $^{60}_{26}Ni$ | **+0.0054234** |
| $^{107}_{47}Ag - (4)^{4}_{2}He \uparrow \xrightarrow{\alpha}$<br>$^{91}_{40}Zr - ^{4}_{2}He \xrightarrow{\alpha}$ | $^{91}_{39}Y$ | $^{91}_{40}Zr$ **absent**<br>$^{87}_{38}Sr$ | +0.010961<br>**+0.005834** |
| $^{109}_{47}Ag - (2)^{4}_{2}He \uparrow \xrightarrow{\alpha}$<br>$^{101}_{44}Ru \xrightarrow{fission} {}^{50}_{22}Ti$<br>$\xrightarrow{fission} {}^{51}_{22}Ti$ | $^{101}_{43}Tc$ | $^{101}_{44}Ru$ **(note 2)**<br>$^{50}_{22}Ti$<br>$^{51}_{23}V$ | **+0.006036**<br>-1.009963 **H**<br>+0.968373 **H** |
| $^{59}_{28}Ni - (1)^{4}_{2}He \xrightarrow{\alpha}$ | $^{55}_{26}Fe$ | $^{55}_{25}Mn$ | **+0.0063016** |
| $^{60}_{28}Ni - (1)^{4}_{2}He \xrightarrow{\alpha}$ | $^{56}_{26}Fe$ | $^{56}_{26}Fe$ | **+0.0067543** |
| $^{58}_{28}Ni - (1)^{4}_{2}He \xrightarrow{\alpha}$ | $^{54}_{26}Fe$ | $^{54}_{26}Fe$ | **+0.006870** |
| $^{61}_{28}Ni - ^{4}_{2}He \xrightarrow{\alpha}$ | $^{57}_{26}Fe$ | | **+0.0069412** |
| $^{61}_{28}Ni - (1)^{4}_{2}He \xrightarrow{\alpha}$ | $^{57}_{26}Fe$ | $^{57}_{26}Fe$ | **+0.0069412** |
| $^{107}_{47}Ag - (2)^{4}_{2}He \uparrow \xrightarrow{\alpha}$<br>$^{99}_{44}Ru \xrightarrow{fission} {}^{49}_{22}Ti$<br>$\xrightarrow{fission} {}^{50}_{22}Ti$ | $^{99}_{43}Tc$ | $^{99}_{44}Ru$ **absent(note 2)**<br>$^{49}_{22}Ti$<br>$^{50}_{22}Ti$ | +0.006366<br>}**-0.007229** |

| | | | |
|---|---|---|---|
| $^{99}_{44}Ru \xrightarrow{fission} {}^{47}_{22}Ti$<br>$\xrightarrow{fission} {}^{52}_{22}Ti$ | | $^{47}_{22}Ti$<br>$^{52}_{22}Ti$ | }-0.007619 |
| $^{107}_{47}Ag - (3)^{4}_{2}He \uparrow \xrightarrow{\alpha}$ | $^{95}_{41}Nb$ | $^{95}_{42}Mo$ | **+0.008554** |
| $^{107}_{47}Ag - (5)^{4}_{2}He \uparrow \xrightarrow{\alpha}$ | $^{87}_{37}Ru$ | $^{87}_{38}Sr$ | **+0.008554** |
| $^{109}_{47}Ag - (3)^{4}_{2}He \uparrow \xrightarrow{\alpha}$<br>$^{97}_{42}Mo \xrightarrow{fission} {}^{45}_{21}Sc$<br>$\xrightarrow{fission} {}^{52}_{21}Sc$ | $^{97}_{41}Nb$ | $^{97}_{42}Mo$ **(note2)**<br>$^{45}_{21}Sc$<br>$^{52}_{24}Cr$ | **+0.009079**<br>-6.983118 **H**<br>+6.984072 **H** |
| $^{109}_{47}Ag - (4)^{4}_{2}He \uparrow \xrightarrow{\alpha}$ | $^{93}_{38}Y$ | $^{93}_{41}Nb$ | **+0.012039** |
| $^{64}_{28}Ni - (2)^{4}_{2}He \xrightarrow{\alpha}$ | $^{56}_{24}Cr$ | $^{56}_{26}Fe$ | **+0.0121778** |
| $^{107}_{47}Ag \xrightarrow{fission}$ | $^{45}_{21}Sc$<br>$^{62}_{26}Fe$ | $^{62}_{28}Ni$ | }**-0.01241** |
| $^{58}_{28}Ni + ^{64}_{28}Ni - (2)^{4}_{2}He \xrightarrow{fusion}$<br>$^{114}_{50}Sn \xrightarrow{fission}$ | $^{114}_{52}Te$ | $\left[^{114}_{50}Sn\right]$<br>$\left[(2)^{57}_{26}Fe\right]$ | +0.044676<br>**+0.012685** |
| $(2)^{60}_{28}Ni + ^{2}_{1}H^{*} \xrightarrow{fusion}$<br>$^{122}_{52}Te \xrightarrow{fission}$ | $^{122}_{57}La$ | $^{118}_{50}Sn, ^{121}_{51}Sb, \left[^{122}_{52}Te\right]$**absent (note 2)**<br>$(2)^{61}_{28}Ni$ | +0.027369<br>**+0.013562** |
| $^{68}_{30}Zn + (1)^{2}_{1}H \xrightarrow{fusion}$ | $^{70}_{31}Ga$ | $^{70}_{30}Zn$ | **-0.013626** |
| $^{61}_{28}Ni + ^{107}_{47}Ag + (2)^{2}_{1}H^{*} \xrightarrow{fusion}$<br>$^{114}_{50}Sn - ^{4}_{2}He \xrightarrow{\alpha}$<br>$^{112}_{50}Sn \xrightarrow{\beta^{+}\beta^{+}}$<br>$^{113}_{49}In - ^{4}_{2}He \xrightarrow{\alpha}$<br>$^{114}_{50}Sn \xrightarrow{fission}$ **(note 2)** | $^{114}_{55}Cs$ | $^{106}_{48}Cd, ^{107}_{47}Ag, ^{108}_{48}Cd, ^{109}_{47}Ag, ^{110}_{48}Cd,$<br>$^{111}_{48}Cd, ^{112}_{50}Sn, ^{113}_{49}In, ^{114}_{50}Sn$<br>$^{110}_{48}Cd$<br>$^{112}_{48}Cd$<br>$^{109}_{47}Ag$<br>$\left[(2)^{57}_{26}Fe\right]$ | -3.95358<br>-1.951766<br>-4.951832<br>**-0.014203** |
| | $^{107}_{48}Cd$ | $^{109}_{47}Ag$ | **-0.014446** |

| | | | |
|---|---|---|---|
| ${}^{58}_{28}Ni-(2){}^{4}_{2}He\xrightarrow{\alpha}$ | ${}^{50}_{24}Cr\xrightarrow{\beta^{+}\beta^{+}}$ **(note 5)** | ${}^{50}_{22}Ti$ | **+0.014654** |
| ${}^{109}_{47}Ag+(1){}^{2}_{1}H\xrightarrow{fusion}$ | ${}^{111}_{48}Cd$ | | **-0.014675** |
| ${}^{65}_{29}Cu+(2){}^{2}_{1}H\xrightarrow{fusion}$ | ${}^{67}_{30}Zn$ | ${}^{67}_{30}Zn$ | **-0.014763** |
| ${}^{61}_{28}Ni-(2){}^{4}_{2}He\xrightarrow{\alpha}$ | ${}^{53}_{24}Cr$ | | **+0.0147999** |
| ${}^{59}_{28}Ni-(2){}^{4}_{2}He\xrightarrow{\alpha}$ | ${}^{51}_{24}Cr$ | ${}^{51}_{23}V$ | **+0.0148193** |
| ${}^{60}_{28}Ni-(3){}^{4}_{2}He\xrightarrow{\alpha}$<br>${}^{48}_{22}Ti+{}^{2}_{1}H^{+}\xrightarrow{fusion}$<br>${}^{50}_{23}V\xrightarrow{\beta^{-}}$<br>$\xrightarrow{\beta^{+}}$ | ${}^{48}_{22}Ti$<br>**absent** | ${}^{50}_{23}V$<br>${}^{50}_{22}Ti$<br>${}^{50}_{24}Cr$ | +0.024969<br>**-0.014889**<br>-0.0172568<br>-0.0160038 |
| ${}^{60}_{28}Ni-(2){}^{4}_{2}He\xrightarrow{\alpha}$ | ${}^{52}_{24}Cr$ | | **+0.0149276** |
| ${}^{60}_{28}Ni+{}^{61}_{28}Ni-{}^{4}_{2}He\xrightarrow{fusion}$ | ${}^{117}_{54}Xe$ | ${}^{116}_{50}Sn,\left[{}^{117}_{50}Sn\right]$ | **+0.015139** |
| ${}^{58}_{28}Ni+{}^{60}_{28}Ni-(2){}^{4}_{2}He\xrightarrow{fusion}$<br>${}^{110}_{48}Cd\xrightarrow{fission}$ | ${}^{110}_{52}Te$ | ${}^{109}_{47}Ag,\left[{}^{110}_{48}Cd\right]$ **absent**<br>$\left[(2){}^{55}_{24}Mn\right]$ | +0.042079<br>**+0.015167** |
| ${}^{66}_{30}Zn+(1){}^{2}_{1}H\xrightarrow{fusion}$ | ${}^{68}_{31}Zn$ | ${}^{68}_{31}Zn$ | **-0.015290** |
| ${}^{61}_{28}Ni+(1){}^{2}_{1}H\xrightarrow{fusion}$ | ${}^{63}_{29}Cu$ | ${}^{63}_{29}Cu$ | **-0.015560** |
| ${}^{67}_{30}Zn+(1){}^{2}_{1}H\xrightarrow{fusion}$ | ${}^{69}_{31}Ga$ | ${}^{69}_{31}Ga$ | **-0.015655** |
| ${}^{109}_{47}Ag-(5){}^{4}_{2}He\uparrow\xrightarrow{\alpha}$ | ${}^{89}_{37}Ru$ | ${}^{89}_{39}Y$ | **+0.015714** |
| ${}^{63}_{29}Cu+(2){}^{2}_{1}H\xrightarrow{fusion}$ | ${}^{65}_{30}Zn$ | ${}^{65}_{29}Cu$ | **-0.015909** |
| ${}^{64}_{28}Ni+(1){}^{2}_{1}H\xrightarrow{fusion}$ | ${}^{66}_{29}Cu$ | ${}^{66}_{30}Zn$ | **-0.016034** |
| ${}^{64}_{28}Ni\xrightarrow{fission}$ | $(2){}^{32}_{14}Si$ | $(2){}^{32}_{16}S$ | **+0.016176** |
| ${}^{59}_{27}Co+(1){}^{2}_{1}H\xrightarrow{fusion}$ | ${}^{61}_{28}Ni$ | | **-0.0162405** |

| | | | |
|---|---|---|---|
| ${}^{66}_{30}Zn+(2){}^{2}_{1}H\xrightarrow{fusion}$<br>${}^{70}_{32}Ge\xrightarrow{fission}$ | ${}^{70}_{32}Ge$<br>**absent** | $(2){}^{35}_{17}Cl\uparrow$ | -0.029969<br>**-0.016531** |
| ${}^{60}_{28}Ni+(1){}^{2}_{1}H\xrightarrow{fusion}$ | ${}^{62}_{29}Cu$ | ${}^{62}_{28}Ni$ | **-0.016543** |
| ${}^{60}_{28}Ni\xrightarrow{fission}$ | $(2){}^{30}_{14}Si$ | | **+0.0167539** |
| $(2){}^{58}_{28}Ni+{}^{2}_{1}H^{+}\xrightarrow{fusion}$ | ${}^{118}_{57}La$ | ${}^{114}_{50}Sn,{}^{116}_{50}Sn,{}^{117}_{50}Sn,\left[{}^{118}_{50}Sn\right]$ | **+0.016815** |
| ${}^{64}_{30}Zn+(1){}^{2}_{1}H\xrightarrow{fusion}$ | ${}^{66}_{31}Ga$ | ${}^{66}_{30}Zn$ | **-0.017210** |
| ${}^{70}_{30}Zn+(1){}^{2}_{1}H\xrightarrow{fusion}$ | ${}^{72}_{31}Ga$ | ${}^{72}_{32}Ge$ | **-0.017345** |
| ${}^{58}_{28}Ni\xrightarrow{fission}$ | $(2){}^{29}_{14}Si$ | | **+0.0176465** |
| ${}^{58}_{28}Ni+(1){}^{2}_{1}H\xrightarrow{fusion}$ | ${}^{60}_{29}Cu$ | ${}^{60}_{28}Ni$ | **-0.018658** |
| ${}^{58}_{28}Ni+{}^{62}_{28}Ni-(2){}^{4}_{2}He\xrightarrow{fusion}$<br>${}^{112}_{50}Sn\xrightarrow{fission}$ | ${}^{112}_{52}Te$ | $\left[{}^{112}_{50}Sn\right]$<br>$\left[(2){}^{56}_{26}Fe\right]$ | +0.046336<br>**+0.019328** |
| ${}^{64}_{28}Ni-(3){}^{4}_{2}He\xrightarrow{\alpha}$ | ${}^{52}_{22}Ti$ | ${}^{52}_{24}Cr$ | **+0.0203510** |
| ${}^{109}_{47}Ag-(6){}^{4}_{2}He\uparrow\xrightarrow{\alpha}$ | ${}^{85}_{35}Br\uparrow^{2.9\,min}$ | ${}^{85}_{37}Rb$ | **+0.022657** |
| ${}^{62}_{28}Ni-(3){}^{4}_{2}He\xrightarrow{\alpha}$ | ${}^{50}_{22}Ti$ | | **+0.0242558** |
| ${}^{61}_{28}Ni-(3){}^{4}_{2}He\xrightarrow{\alpha}$ | ${}^{49}_{22}Ti$ | ${}^{49}_{22}Ti$ | **+0.0246237** |
| ${}^{107}_{47}Ag\xrightarrow{fission}$ | ${}^{53}_{22}Ti$<br>${}^{54}_{25}Mn$ | ${}^{53}_{24}Cr$<br>$\xrightarrow{EC(99.99\%)}{}^{54}_{24}Cr$<br>$\xrightarrow{\beta^{+}(3x10^{-4}\%)}{}^{54}_{26}Fe$<br>$\xrightarrow{\beta^{+}(6x10^{-?}\%)}{}^{54}_{24}Cr$ | -0.0255672<br>**-0.0248371**<br>-0.0255672 |
| ${}^{58}_{28}Ni-(3){}^{4}_{2}He\xrightarrow{\alpha}$ | ${}^{46}_{22}Ti$ | ${}^{46}_{22}Ti$ | **+0.0250984** |
| ${}^{59}_{28}Ni-(3){}^{4}_{2}He\xrightarrow{\alpha}$ | ${}^{47}_{22}Ti$ | ${}^{47}_{22}Ti$ | **+0.0252261** |

| | | | |
|---|---|---|---|
| $^{107}_{47}Ag \xrightarrow{fission}$ | $^{53}_{23}V$<br>$^{54}_{24}Cr$ | $^{53}_{24}Cr$ | **-0.0255672** |
| $^{107}_{47}Ag \xrightarrow{fission}$ | $^{54}_{23}V$<br>$^{53}_{24}Cr$ | $^{54}_{24}Cr$ | **-0.0255672** |
| $^{107}_{47}Ag \xrightarrow{fission}$ | $^{54}_{22}Ti$<br>$^{53}_{25}Mn$ | $^{54}_{24}Cr$<br>$^{53}_{24}Cr$ | **-0.0255672** |
| $^{107}_{47}Ag - (6)^{4}_{2}He \uparrow \xrightarrow{\alpha}$ | $^{83}_{35}Br \uparrow^{2.9 min}$ | $^{85}_{37}Rb$ | **+0.025632** |
| $^{107}_{47}Ag \xrightarrow{fission}$ | $^{51}_{23}V$<br>$^{56}_{24}Cr$ | $^{56}_{26}Fe$ | **-0.0262000** |
| $^{107}_{47}Ag \xrightarrow{fission}$ | $^{51}_{22}Ti$<br>$^{56}_{25}Mn$ | $^{51}_{23}V$<br>$^{56}_{26}Fe$ | **-0.0262000** |
| $^{107}_{47}Ag \xrightarrow{fission}$ | $^{52}_{23}V$<br>$^{55}_{24}Cr$ | $^{52}_{24}Cr$<br>$^{55}_{25}Mn$ | **-0.0265438** |
| $^{107}_{47}Ag \xrightarrow{fission}$ | $^{55}_{23}V$<br>$^{52}_{24}Cr$ | $^{55}_{25}Mn$ | **-0.0265438** |
| $^{107}_{47}Ag \xrightarrow{fission}$ | $^{52}_{22}Ti$<br>$^{55}_{25}Mn$ | $^{52}_{24}Cr$ | **-0.0265438** |
| $^{107}_{47}Ag \xrightarrow{fission}$ | $^{55}_{22}Ti$<br>$^{52}_{25}Mn$ | $^{55}_{25}Mn$<br>$^{52}_{24}Cr$ | **-0.0265438** |
| $^{109}_{47}Ag \xrightarrow{fission}$ | $^{50}_{20}Ca$<br>$^{59}_{27}Co$ | $^{50}_{22}Ti$ | **-0.0267658** |
| $^{109}_{47}Ag \xrightarrow{fission}$ | $^{55}_{22}Ti$<br>$^{54}_{25}Mn$ | $^{55}_{25}Mn$<br>$\xrightarrow{EC(99.99\%)} {}^{54}_{24}Cr$<br>$\xrightarrow{\beta^-(3x10^{-4}\%)} {}^{54}_{26}Fe$ | +0.02782582<br>**+0.0270965** |

| | | | |
|---|---|---|---|
| $^{61}_{28}Ni + (2)^{2}_{1}H \xrightarrow{fusion}$ | $^{65}_{30}Zn$ | $^{65}_{29}Cu$ | **-0.031470** |
| $^{59}_{27}Co + (2)^{2}_{1}H \xrightarrow{fusion}$ | $^{63}_{29}Cu$ | | **-0.0318005** |
| $^{70}_{30}Zn + (2)^{2}_{1}H \xrightarrow{fusion}$ | $^{74}_{32}Ge$ | $^{74}_{32}Ge$ | **-0.032344** |
| $^{64}_{30}Zn + (2)^{2}_{1}H \xrightarrow{fusion}$ | $^{68}_{32}Ge$ | $^{68}_{30}Zn$ | **-0.032501** |
| $^{64}_{28}Ni + (3)^{2}_{1}H \xrightarrow{fusion}$<br>$^{70}_{31}Ge \xrightarrow{fission}$ | $^{70}_{31}Ga$ | $\xrightarrow{EC(4\%)} {}^{70}_{30}Zn$<br>$\xrightarrow{\beta^-(99.6\%)} {}^{70}_{32}Ge$ **absent**<br>$(2)^{35}_{17}Cl \uparrow$ | -0.044952<br>-0.046022<br>**-0.032565** |
| $(3)^{64}_{28}Ni \xrightarrow{fusion}$<br>$^{192}_{84}Pt \xrightarrow{fission}$ | $^{192}_{84}Po$<br>$(2)^{96}_{39}Y$<br>$(2)^{97}_{39}Y$ | $^{176}_{72}Hf, ^{184}_{76}Os, ^{188}_{76}Os, [^{192}_{78}Pt]$<br>$[(2)^{96}_{40}Zr]$<br>$[(2)^{97}_{42}Mo]$ | +0.207437<br>**+0.032648**<br>+2.028145 |
| $^{64}_{30}Zn + (3)^{2}_{1}H \xrightarrow{fusion}$<br>$^{70}_{32}Ge \xrightarrow{fission}$ | $^{70}_{33}As$ | $^{70}_{32}Ge$ **absent**<br>$(2)^{35}_{17}Cl \uparrow$ | -0.047200<br>**-0.033743** |
| $^{58}_{28}Ni + (2)^{2}_{1}H \xrightarrow{fusion}$ | $^{62}_{30}Zn$ | $^{62}_{28}Ni$ | **-0.035201** |
| $^{68}_{30}Zn + (3)^{2}_{1}H \xrightarrow{fusion}$<br>$^{74}_{34}Se \xrightarrow{fission}$ | $^{74}_{33}As$ | $\xrightarrow{\beta^+(66\%)} {}^{74}_{32}Ge$<br>$\xrightarrow{\beta^-(34\%)} {}^{74}_{34}Se$<br>$(2)^{37}_{17}Cl \uparrow$ | -0.045971<br>-0.044672<br>**-0.035344** |
| $^{109}_{47}Ag - (8)^{4}_{2}He \uparrow \xrightarrow{\alpha}$ | $^{75}_{31}Ga$ | $^{77}_{34}Se$ | **+0,035988** |
| $^{107}_{47}Ag - (8)^{4}_{2}He \uparrow \xrightarrow{\alpha}$ | $^{71}_{29}Cu$ | $^{75}_{33}As$ | **+0.037325** |
| $^{58}_{28}Ni + ^{61}_{28}Ni - (2)^{4}_{2}He \xrightarrow{fusion}$ | $^{111}_{52}Te$ | $^{110}_{48}Cd, [^{111}_{48}Cd]$ | **+0.042985** |
| $^{107}_{47}Ag + (3)^{2}_{1}H \xrightarrow{fusion}$ | $^{113}_{50}Sn$ | $^{113}_{49}In$ | **-0.043343** |
| $^{60}_{28}Ni + ^{62}_{28}Ni - ^{4}_{2}He \xrightarrow{fusion}$ | $^{118}_{54}Xe$ | $[^{118}_{50}Sn]$ | **+0.045074** |

| | | | |
|---|---|---|---|
| | | $\xrightarrow{\beta^+(6x10^{-7}s)} {}^{54}_{24}Cr$ | +0.02782582 |
| ${}^{62}_{28}Ni - (1)^4_2He \xrightarrow{\alpha}$ | ${}^{58}_{26}Fe \xrightarrow{fission}$ | $(2)^{29}_{14}Si$ | **+0.0272476** |
| ${}^{109}_{47}Ag + (2)^2_1H \xrightarrow{fusion}$ | ${}^{113}_{49}In$ | | **-0.028897** |
| ${}^{107}_{47}Ag + (2)^2_1H \xrightarrow{fusion}$ | ${}^{111}_{49}In$ | ${}^{111}_{48}Cd$ | **-0.029121** |
| ${}^{109}_{47}Ag \xrightarrow{fission}$ | ${}^{53}_{22}Ti$<br>${}^{56}_{25}Mn$ | ${}^{53}_{24}Cr$<br>${}^{56}_{26}Fe$ | } **-0.0291651** |
| ${}^{109}_{47}Ag \xrightarrow{fission}$ | ${}^{56}_{22}Ti$<br>${}^{53}_{25}Mn$ | ${}^{56}_{26}Fe$<br>${}^{53}_{24}Cr$ | } **-0.0291651** |
| ${}^{109}_{47}Ag \xrightarrow{fission}$ | ${}^{53}_{23}V$<br>${}^{56}_{24}Cr$ | ${}^{53}_{24}Cr$<br>${}^{56}_{26}Fe$ | } **-0.0291651** |
| ${}^{109}_{47}Ag \xrightarrow{fission}$ | ${}^{56}_{23}V$<br>${}^{53}_{24}Cr$ | ${}^{56}_{26}Fe$ | } **-0.0291651** |
| ${}^{107}_{47}Ag - (7)^4_2He \uparrow \xrightarrow{\alpha}$ | ${}^{79}_{33}As$ | ${}^{79}_{35}Br \uparrow$ | **+0.029649** |
| ${}^{109}_{47}Ag - (7)^4_2He \uparrow \xrightarrow{\alpha}$ | ${}^{81}_{33}As$ | ${}^{81}_{35}Br \uparrow$ | **+0.029761** |
| ${}^{109}_{47}Ag \xrightarrow{fission}$ | ${}^{54}_{22}Ti$<br>${}^{55}_{25}Mn$ | ${}^{54}_{24}Cr$ | } **-0.0298266** |
| ${}^{60}_{28}Ni + (2)^2_1H \xrightarrow{fusion}$ | ${}^{64}_{30}Zn$ | ${}^{64}_{30}Zn$ | **-0.029847** |
| ${}^{65}_{29}Cu + (3)^2_1H \xrightarrow{fusion}$ | ${}^{69}_{31}Ga$ | ${}^{69}_{31}Ga$ | **-0.030418** |
| ${}^{62}_{28}Ni + (2)^2_1H \xrightarrow{fusion}$ | ${}^{66}_{30}Zn$ | ${}^{66}_{30}Zn$ | **-0.030515** |
| ${}^{67}_{30}Zn + (2)^2_1H \xrightarrow{fusion}$ | ${}^{71}_{32}Ge$ | ${}^{71}_{31}Ga$ | **-0.030629** |
| ${}^{63}_{29}Cu + (3)^2_1H \xrightarrow{fusion}$ | ${}^{67}_{31}Ga$ | ${}^{67}_{30}Zn$ | **-0.030673** |
| ${}^{68}_{30}Zn + (2)^2_1H \xrightarrow{fusion}$ | ${}^{72}_{32}Ge$ | ${}^{72}_{32}Ge$ | **-0.030971** |
| ${}^{64}_{28}Ni + (2)^2_1H \xrightarrow{fusion}$ | ${}^{68}_{30}Zn$ | ${}^{68}_{30}Zn$ | **-0.031325** |

| | | | |
|---|---|---|---|
| ${}^{115}_{51}Sb - {}^4_2He \xrightarrow{\alpha}$ | | | |
| ${}^{109}_{47}Ag + (4)^2_1H \xrightarrow{fusion}$ | ${}^{117}_{51}Sb$ | ${}^{117}_{50}Sn$ | **-0.0582061** |
| ${}^{66}_{30}Zn + (5)^2_1H \xrightarrow{fusion}$<br>${}^{76}_{34}Se \xrightarrow{fission}$<br>$(2)^{38}_{17}Cl \uparrow^{37min} \xrightarrow{\beta^-}$ | ${}^{76}_{35}Br \uparrow^{16hrs}$ | ${}^{76}_{34}Se$<br>$(2)^{38}_{17}Cl \uparrow^{37min}$<br>$(2)^{40}_{17}Ar \uparrow$ | -0.077328<br>**-0.060521**<br>-0.071077 |
| ${}^{65}_{29}Cu + (5)^2_1H \xrightarrow{fusion}$ | ${}^{73}_{33}As$ | ${}^{73}_{32}Ge$ | **-0.060736** |
| ${}^{63}_{29}Cu + (5)^2_1H \xrightarrow{fusion}$ | ${}^{71}_{33}As$ | ${}^{71}_{31}Ga$ | **-0.061302** |
| ${}^{67}_{30}Zn + (4)^2_1H \xrightarrow{fusion}$ | ${}^{75}_{34}Se$ | ${}^{75}_{33}As$ | **-0.061636** |
| ${}^{61}_{28}Ni + (4)^2_1H \xrightarrow{fusion}$ | ${}^{69}_{32}Ge$ | ${}^{69}_{31}Ga$ | **-0.061889** |
| ${}^{68}_{30}Zn + (4)^2_1H \xrightarrow{fusion}$ | ${}^{76}_{34}Se$ | | **-0.062036** |
| ${}^{64}_{28}Ni + (4)^2_1H \xrightarrow{fusion}$ | ${}^{72}_{32}Ge$ | | **-0.062297** |
| ${}^{60}_{28}Ni + (4)^2_1H \xrightarrow{fusion}$ | ${}^{68}_{32}Ge$ | ${}^{68}_{30}Zn$ | **-0.062349** |
| ${}^{59}_{27}Co + (4)^2_1H \xrightarrow{fusion}$ | ${}^{67}_{31}Ga$ | ${}^{67}_{30}Zn$ | **-0.0624738** |
| ${}^{64}_{30}Zn + (4)^2_1H \xrightarrow{fusion}$ | ${}^{72}_{34}Se$ | ${}^{72}_{32}Ge$ | **-0.063472** |
| ${}^{60}_{28}Ni + (5)^2_1H \xrightarrow{fusion}$<br>${}^{70}_{32}Ge \xrightarrow{fission}$ | ${}^{70}_{33}As$ | ${}^{70}_{32}Ge$<br>$(2)^{35}_{16}S \xrightarrow{\beta^-} (2)^{35}_{17}Cl \uparrow$ | -0.377049<br>**-0.063579** |
| ${}^{58}_{28}Ni + (4)^2_1H \xrightarrow{fusion}$ | ${}^{66}_{32}Ge$ | ${}^{66}_{30}Zn$ | **-0.065716** |
| ${}^{64}_{28}Ni + (5)^2_1H \xrightarrow{fusion}$<br><br>${}^{74}_{34}Se \xrightarrow{fission}$ | ${}^{74}_{33}As \xrightarrow{17day}$ | $\xrightarrow{\beta^+(66\%)} {}^{74}_{32}Ge$<br>$\xrightarrow{\beta^-(34\%)} {}^{74}_{34}Se$ **absent**<br>$(2)^{37}_{17}Cl$ | -0.077293<br>-0.075994<br>**-0.066666** |
| ${}^{64}_{30}Zn + (5)^2_1H \xrightarrow{fusion}$<br>${}^{74}_{34}Se \xrightarrow{fission}$ | ${}^{74}_{35}Br \uparrow^{25min}$ | ${}^{74}_{34}Se$ **absent**<br>$(2)^{37}_{17}Cl \uparrow$ | -0.077174<br>**-0.067825** |
| ${}^{109}_{47}Ag + (5)^2_1H \xrightarrow{fusion}$ | ${}^{119}_{52}Te$ | ${}^{119}_{50}Sn$ | **-0.0719517** |
| ${}^{64}_{28}Ni + {}^{107}_{47}Ag + (2)^2_1H^* \xrightarrow{fusion}$ | ${}^{172}_{77}Ir$ | ${}^{164}_{68}Er, {}^{168}_{70}Yb, \left[{}^{172}_{70}Yb\right]$ | **+0.0720249** |

| | | | |
|---|---|---|---|
| ${}^{107}_{47}Ag+(5)^2_1H \xrightarrow{fusion}$ | ${}^{117}_{52}Te$ | ${}^{117}_{50}Sn$ | **-0.0726527** |
| ${}^{68}_{30}Zn+(5)^2_1H \xrightarrow{fusion}$ | ${}^{78}_{35}Br \uparrow^{6min}$ | $\xrightarrow{\beta^+(99+\%)} {}^{78}_{34}Se$<br>$\xrightarrow{\beta^-(.01\%)} {}^{78}_{36}Kr \uparrow$ | -0.078043<br>**-0.074988** |
| ${}^{63}_{29}Cu+(6)^2_1H \xrightarrow{fusion}$ | ${}^{73}_{34}Se$ | ${}^{73}_{32}Ge$ | **-0.076646** |
| ${}^{65}_{29}Cu+(6)^2_1H \xrightarrow{fusion}$ | ${}^{75}_{34}Se$ | ${}^{75}_{33}As$ | **-0.076700** |
| ${}^{62}_{28}Ni+(5)^2_1H \xrightarrow{fusion}$ | ${}^{72}_{33}As$ | ${}^{72}_{31}Ge$ | **-0.076778** |
| ${}^{58}_{28}Ni+{}^{107}_{47}Ag+{}^2_1H^+ \xrightarrow{fusion}$<br>${}^{167}_{68}Er \xrightarrow{n}$ | ${}^{167}_{76}Os$ | ${}^{143}_{60}Nd, {}^{147}_{62}Sm, {}^{155}_{64}Gd, {}^{159}_{65}Tb, {}^{163}_{66}Dy, \left[{}^{167}_{68}Er\right]$<br>$\left[{}^{163}_{66}Dy\right]$ | +0.0775065<br>**+0.0767937** |
| ${}^{61}_{28}Ni+(5)^2_1H \xrightarrow{fusion}$ | ${}^{71}_{33}As$ | ${}^{71}_{33}Ga$ | **-0.076863** |
| ${}^{67}_{30}Zn+(5)^2_1H \xrightarrow{fusion}$ | ${}^{77}_{35}Br \uparrow^{57hrs}$ | ${}^{77}_{34}Se$ | **-0.077721** |
| ${}^{59}_{27}Co+(5)^2_1H \xrightarrow{fusion}$ | ${}^{69}_{32}Ge$ | ${}^{69}_{31}Ga$ | **-0.0781291** |
| ${}^{109}_{47}Ag+(3)^2_1H \xrightarrow{fusion}$<br>${}^{115}_{50}Sn \xrightarrow{fission} {}^{56}_{25}Mn$<br>and | ${}^{115}_{50}Sn$<br>**note 2** | ${}^{56}_{26}Fe$<br>${}^{59}_{27}Co$ | } **-0.078924** |
| ${}^{70}_{30}Zn+(5)^2_1H \xrightarrow{fusion}$ | ${}^{80}_{35}Br \uparrow^{18min}$ | $\xrightarrow{\beta^-(92\%)} {}^{80}_{36}Kr \uparrow$<br>$\xrightarrow{\beta^+(8\%)} {}^{80}_{34}Se$ | -0.079449<br>**-0.079306** |
| ${}^{58}_{28}Ni+(5)^2_1H \xrightarrow{fusion}$ | ${}^{68}_{33}As$ | ${}^{68}_{30}Zn$ | **-0.081007** |
| ${}^{62}_{28}Ni+(6)^2_1H \xrightarrow{fusion}$ | ${}^{74}_{34}Se \xrightarrow{fission}$ | ${}^{37}_{17}Cl \uparrow$ | **-0.081150** |
| ${}^{58}_{28}Ni+(6)^2_1H \xrightarrow{fusion}$<br>${}^{70}_{32}Ge \xrightarrow{fission}$ | ${}^{70}_{34}Se$ | ${}^{70}_{32}Ge$ **absent**<br>$(2){}^{35}_{16}S \xrightarrow{\beta^-} (2){}^{35}_{17}Cl \uparrow$ | -0.095706<br>**-0.082248** |
| ${}^{107}_{47}Ag+(6)^2_1H \xrightarrow{fusion}$ | ${}^{119}_{53}I \uparrow^{19min}$ | ${}^{119}_{50}Sn$ | **-0.0838612** |
| ${}^{109}_{47}Ag+(6)^2_1H \xrightarrow{fusion}$ | ${}^{121}_{53}I \uparrow^{2hrs}$ | ${}^{121}_{51}Sb$ | **-0.0855455** |

| | | | |
|---|---|---|---|
| ${}^{65}_{29}Cu+(4)^2_1H \xrightarrow{fusion}$ | ${}^{71}_{32}Ge$ | ${}^{71}_{31}Ga$ | **-0.045392** |
| ${}^{70}_{30}Zn+(4)^2_1H \xrightarrow{fusion}$<br>$(2){}^{39}_{17}Cl \uparrow^{59min} \xrightarrow{\beta^-}$ | ${}^{78}_{34}Se \xrightarrow{fission}$<br>**absent** | $(2){}^{39}_{17}Cl \uparrow^{59min}$<br>$(2){}^{39}_{18}Ar \uparrow^{269yr}$ | -0.064417<br>**-0.045710**<br>-0.053099 |
| ${}^{62}_{28}Ni+(3)^2_1H \xrightarrow{fusion}$ | ${}^{68}_{31}Ge$ | ${}^{68}_{30}Zn$ | **-0.045806** |
| ${}^{67}_{30}Zn+(3)^2_1H \xrightarrow{fusion}$ | ${}^{73}_{33}As$ | ${}^{73}_{32}Ge$ | **-0.045973** |
| ${}^{70}_{30}Zn+(3)^2_1H \xrightarrow{fusion}$<br>${}^{76}_{32}Ge \xrightarrow{\beta^-\beta^-}$ | ${}^{76}_{33}As$ | $\xrightarrow{\beta^-(99+\%)} {}^{76}_{34}Se$<br>$\xrightarrow{EC(.02\%)} {}^{76}_{32}Ge$<br>${}^{76}_{34}Se$ | -0.048411<br>**-0.046222**<br>-0.048411 |
| ${}^{61}_{28}Ni+(3)^2_1H \xrightarrow{fusion}$ | ${}^{67}_{31}Ga$ | ${}^{67}_{30}Zn$ | **-0.046234** |
| ${}^{66}_{30}Zn+(3)^2_1H \xrightarrow{fusion}$ | ${}^{72}_{33}As$ | ${}^{72}_{32}Ge$ | **-0.046262** |
| ${}^{63}_{29}Cu+(4)^2_1H \xrightarrow{fusion}$ | ${}^{69}_{32}Ge$ | ${}^{69}_{31}Ga$ | **-0.046328** |
| ${}^{62}_{28}Ni+(4)^2_1H \xrightarrow{fusion}$ | ${}^{70}_{32}Ge \xrightarrow{fission}$ | ${}^{35}_{17}Cl \uparrow$ | **-0.047048** |
| ${}^{60}_{28}Ni+(3)^2_1H \xrightarrow{fusion}$ | ${}^{66}_{31}Ga$ | ${}^{66}_{30}Zn$ | **-0.047058** |
| $(2){}^{64}_{28}Ni \xrightarrow{fusion}$ | ${}^{128}_{56}Ba$ | $\left[{}^{128}_{54}Xe\right]$ | **-0.0475993** |
| ${}^{59}_{27}Co+(3)^2_1H \xrightarrow{fusion}$ | ${}^{65}_{30}Zn$ | ${}^{65}_{29}Cu$ | **-0.0477101** |
| ${}^{58}_{28}Ni+(3)^2_1H \xrightarrow{fusion}$ | ${}^{64}_{31}Ga$ | ${}^{64}_{30}Zn$ | **-0.048506** |
| $(2){}^{62}_{28}Ni \xrightarrow{fusion}$ | ${}^{124}_{56}Ba$ | $\left[{}^{124}_{54}Xe\right]$ | **+0.0492028** |
| ${}^{66}_{30}Zn+(4)^2_1H \xrightarrow{fusion}$<br>${}^{74}_{34}Se \xrightarrow{fission}$ | ${}^{74}_{34}Se$ | $(2){}^{37}_{17}Cl$ | -0.059963<br>**-0.050634** |
| ${}^{107}_{47}Ag+(4)^2_1H \xrightarrow{fusion}$<br>${}^{115}_{51}Sb \xrightarrow{\beta^+} {}^{115}_{50}Sn$ **absent(**<br>**note1)** | ${}^{115}_{51}Sb$ | ${}^{111}_{49}Cd$ | -0.0549061<br>**-0.0547226** |

| | | | |
|---|---|---|---|
| $^{66}_{30}Zn + (6)^2_1H \xrightarrow{fusion}$ | $^{78}_{36}Kr\uparrow$ | $^{78}_{36}Kr\uparrow$ | **-0.090277** |
| $^{64}_{28}Ni + (7)^2_1H \xrightarrow{fusion}$<br>$^{78}_{34}Se \xrightarrow{fission}$ | $^{78}_{35}Br\uparrow^{6\,min}$ | $\xrightarrow{\beta^+(99.9\%)} {}^{78}_{34}Se$ **absent**<br>$\xrightarrow{\beta^-(0.01\%)} {}^{78}_{36}Kr\uparrow$<br>$^{39}_{18}Ar\uparrow^{269\,y}$ | -0.109369<br>-0.106313<br>**+0.090805** |
| $^{61}_{28}Ni + (6)^2_1H \xrightarrow{fusion}$ | $^{73}_{34}Se$ | $^{73}_{32}Ge$ | **-0.092207** |
| $^{65}_{29}Cu + (7)^2_1H \xrightarrow{fusion}$ | $^{77}_{35}Br\uparrow^{57\,hrs}$ | $^{77}_{34}Se$ | **-0.092484** |
| $^{63}_{29}Cu + (7)^2_1H \xrightarrow{fusion}$ | $^{75}_{35}Br\uparrow^{96\,min}$ | $^{75}_{33}As$ | **-0.092610** |
| $^{68}_{30}Zn + (6)^2_1H \xrightarrow{fusion}$ | $^{80}_{36}Kr\uparrow$ | $^{80}_{36}Kr\uparrow$ | **-0.093074** |
| $^{59}_{27}Co + (6)^2_1H \xrightarrow{fusion}$ | $^{71}_{33}As$ | $^{71}_{31}Ga$ | **-0.0931029** |
| $^{60}_{28}Ni + (6)^2_1H \xrightarrow{fusion}$ | $^{72}_{34}Se$ | $^{72}_{32}Ge$ | **-0.093321** |
| $^{64}_{28}Ni + (6)^2_1H \xrightarrow{fusion}$ | $^{76}_{34}Se$ | | **-0.093363** |
| $^{67}_{30}Zn + (6)^2_1H \xrightarrow{fusion}$ | $^{79}_{36}Kr\uparrow^{35\,hrs}$ | $^{79}_{35}Br\uparrow$ | **-0.093399** |
| $^{64}_{30}Zn + (6)^2_1H \xrightarrow{fusion}$ | $^{76}_{36}Kr\uparrow^{14.8\,hrs}$ | $^{76}_{34}Se$ | **-0.094537** |
| $^{70}_{30}Zn + (6)^2_1H \xrightarrow{fusion}$ | $^{82}_{36}Kr\uparrow$ | | **-0.096444** |
| $^{109}_{47}Ag + (7)^2_1H \xrightarrow{fusion}$<br>$^{123}_{52}Te \xrightarrow{\beta^+}$<br>$^{123}_{52}Te - {}^4_2He \xrightarrow{\alpha}$ | $^{123}_{54}Xe\uparrow^{2.1\,hr}$ | $^{123}_{52}Te$ **absent(note 4)**<br>$^{123}_{51}Sb$ **absent**<br>$^{119}_{50}Sn$ | **-0.0991943**<br>**-0.0992503**<br>**-0.0975531** |
| $^{60}_{28}Ni + (7)^2_1H \xrightarrow{fusion}$<br>$^{74}_{34}Se \xrightarrow{fission}$ | $^{74}_{35}Br\uparrow^{25\,min}$ | $^{74}_{34}Se$ **absent**<br>$(2)^{37}_{17}Cl\uparrow$ | -0.107011<br>**-0.097682** |
| $^{107}_{47}Ag + (7)^2_1H \xrightarrow{fusion}$ | $^{121}_{54}Xe\uparrow^{40\,min}$ | $^{121}_{51}Sb$ | **-0.0999920** |

| | | | |
|---|---|---|---|
| $^{62}_{28}Ni + (8)^2_1H \xrightarrow{fusion}$ | $^{78}_{36}Kr\uparrow$ | $^{78}_{36}Kr\uparrow$ | **-0.120794** |
| $^{61}_{28}Ni + (8)^2_1H \xrightarrow{fusion}$ | $^{77}_{36}Kr\uparrow^{74\,min}$ | $^{77}_{35}Se$ | **-0.123956** |
| $^{63}_{29}Cu + (9)^2_1H \xrightarrow{fusion}$ | $^{79}_{37}Rb$ | $^{79}_{35}Br\uparrow$ | **-0.124072** |
| $^{60}_{28}Ni + (8)^2_1H \xrightarrow{fusion}$ | $^{76}_{36}Kr\uparrow^{76\,hr}$ | $^{76}_{34}Se$ | **-0.124386** |
| $^{64}_{28}Ni + (8)^2_1H \xrightarrow{fusion}$ | $^{80}_{36}Kr\uparrow^{stable}$ | $^{80}_{36}Kr\uparrow$ | **-0.124401** |
| $^{59}_{27}Co + (8)^2_1H \xrightarrow{fusion}$ | $^{75}_{35}Br\uparrow^{96\,min}$ | $^{75}_{33}As$ | **-0.1244108** |
| $^{66}_{30}Zn + (8)^2_1H \xrightarrow{fusion}$ | $^{82}_{38}Sr$ | $^{82}_{36}Kr\uparrow$ | **-0.125362** |
| $^{64}_{30}Zn + (8)^2_1H \xrightarrow{fusion}$ | $^{80}_{38}Sr$ | $^{80}_{36}Kr\uparrow$ | **-0.125575** |
| $^{67}_{30}Zn + (8)^2_1H \xrightarrow{fusion}$ | $^{83}_{38}Sr$ | $^{83}_{36}Kr\uparrow$ | **-0.125803** |
| $^{60}_{28}Ni + (9)^2_1H \xrightarrow{fusion}$ | $^{78}_{37}Rb$ | $^{78}_{36}Kr\uparrow$ | **-0.137337** |
| $^{62}_{28}Ni + (9)^2_1H \xrightarrow{fusion}$ | $^{80}_{37}Rb$ | $^{80}_{36}Kr\uparrow$ | **-0.138882** |
| $^{61}_{28}Ni + (9)^2_1H \xrightarrow{fusion}$ | $^{79}_{37}Rb$ | $^{79}_{35}Br\uparrow$ | **-0.139634** |
| $^{59}_{27}Co + (9)^2_1H \xrightarrow{fusion}$ | $^{77}_{36}Kr\uparrow^{4\,min}$ | $^{77}_{35}Br\uparrow^{57\,hr} \xrightarrow{\beta^+} {}^{77}_{34}Se$ | **-0.1401948** |
| $^{65}_{29}Cu + (1)^2_1H \xrightarrow{fusion}$ | $^{81}_{38}Sr$ | $^{81}_{35}Br\uparrow$ | **-0.140220** |
| $^{58}_{28}Ni + (9)^2_1H \xrightarrow{fusion}$ | $^{76}_{37}Rb$ | $^{76}_{34}Se$<br>$^{76}_{32}Ge$ | -0.143045<br>**-0.140856** |
| $^{64}_{28}Ni + (9)^2_1H \xrightarrow{fusion}$ | $^{80}_{37}Rb$ | $^{82}_{36}Kr\uparrow$ | **-0.141398** |
| $^{60}_{28}Ni + (10)^2_1H \xrightarrow{fusion}$ | $^{80}_{38}Sr$ | $^{80}_{36}Kr\uparrow$ | **-0.155425** |
| $^{61}_{28}Ni + (10)^2_1H \xrightarrow{fusion}$ | $^{81}_{32}Sr$ | $^{81}_{35}Br\uparrow$ | **-0.155783** |
| $^{62}_{28}Ni + (10)^2_1H \xrightarrow{fusion}$ | $^{84}_{38}Sr$ | $^{82}_{36}Kr$ | **-0.155879** |
| $^{58}_{28}Ni + (10)^2_1H \xrightarrow{fusion}$ | $^{78}_{38}Sr$ | $^{78}_{36}Kr\uparrow$ | **-0.155995** |

| | | | |
|---|---|---|---|
| ${}^{70}_{30}Zn+(7)^{2}_{1}H\xrightarrow{fusion}$<br><br>${}^{64}_{38}Sr\xrightarrow{fusion}$<br>${}^{64}_{38}Sr\xrightarrow{fusion}$<br>${}^{45}_{21}Sc+{}^{39}_{17}Cl\uparrow^{56\,min}\rightarrow$ | ${}^{84}_{37}Rb$ | $\xrightarrow{\beta^{+}(96\%)}{}^{84}_{36}Kr\uparrow$<br>$\xrightarrow{\beta^{+}(4\%)}{}^{84}_{38}Sr$ **absent**<br>$(2){}^{42}_{20}Ca$ **absent**<br>${}^{45}_{21}Sc+{}^{39}_{17}Cl\uparrow^{56\,min}$<br>${}^{45}_{21}Sc+{}^{19}_{9}Fl\uparrow+{}^{20}_{10}Ne\uparrow$ | -0.112524<br>-0.110606<br>-0.106795<br>**-0.100111**<br>-0.577276 |
| ${}^{64}_{30}Zn+(7)^{2}_{1}H\xrightarrow{fusion}$ | ${}^{78}_{37}Rb$ | ${}^{78}_{36}Kr\uparrow$ | **-0.107488** |
| ${}^{62}_{28}Ni+(7)^{2}_{1}H\xrightarrow{fusion}$ | ${}^{76}_{35}Br\uparrow^{16\,hrs}$ | ${}^{76}_{34}Se$ | **-0.107843** |
| | ${}^{79}_{36}Kr\uparrow^{35\,hrs}$ | ${}^{79}_{35}Br\uparrow$ | **-0.108163** |
| ${}^{61}_{28}Ni+(7)^{2}_{1}H\xrightarrow{fusion}$ | ${}^{75}_{35}Br\uparrow^{96\,min}$ | ${}^{75}_{33}As$ | **-0.108171** |
| ${}^{66}_{30}Zn+(7)^{2}_{1}H\xrightarrow{fusion}$ | ${}^{80}_{37}Rb$ | ${}^{80}_{36}Kr\uparrow$ | **-0.108365** |
| ${}^{63}_{29}Cu+(8)^{2}_{1}H\xrightarrow{fusion}$ | ${}^{77}_{36}Kr\uparrow^{74\,min}$ | ${}^{77}_{34}Se$ | **-0.108394** |
| ${}^{59}_{27}Co+(7)^{2}_{1}H\xrightarrow{fusion}$ | ${}^{73}_{34}As$ | ${}^{73}_{32}Ge$ | **-0.1084468** |
| ${}^{67}_{30}Zn+(7)^{2}_{1}H\xrightarrow{fusion}$ | ${}^{81}_{37}Rb$ | ${}^{81}_{36}Kr\uparrow$ | **-0.109246** |
| ${}^{68}_{30}Zn+(7)^{2}_{1}H\xrightarrow{fusion}$ | ${}^{82}_{37}Kr\uparrow$ | ${}^{82}_{37}Kr\uparrow$ | **-0.110071** |
| ${}^{58}_{28}Ni+(7)^{2}_{1}H\xrightarrow{fusion}$ | ${}^{72}_{35}Br\uparrow^{79\,sec}$ | ${}^{72}_{32}Ge$ | **-0.111979** |
| ${}^{107}_{47}Ag+(8)^{2}_{1}H\xrightarrow{fusion}$<br>${}^{123}_{52}Te\xrightarrow{\beta^{+}}$<br>${}^{123}_{52}Te-{}^{4}_{2}He\xrightarrow{\alpha}$ | ${}^{123}_{55}Cs$ | ${}^{123}_{52}Te$ **absent(note 4)**<br>${}^{123}_{51}Sb$ **absent**<br>${}^{119}_{50}Sn$ | -0.113640<br>-0.113697<br>**-0.111999** |
| ${}^{109}_{49}Ag+(8)^{2}_{1}H\xrightarrow{fusion}$ | ${}^{125}_{55}Cs$ | ${}^{125}_{53}I\uparrow^{59\,days}\xrightarrow{EC}{}^{125}_{52}Te$ | **-0.1131336** |
| ${}^{58}_{28}Ni+(8)^{2}_{1}H\xrightarrow{fusion}$<br>${}^{74}_{34}Se\xrightarrow{fission}$ | ${}^{74}_{36}Kr\uparrow^{11\,min}$ | ${}^{74}_{34}Se$ **absent**<br>$(2){}^{37}_{17}Cl\uparrow$ | -0.125678<br>**-0.116351** |

www.ingramcontent.com/pod-product-compliance
Lightning Source LLC
Chambersburg PA
CBHW080626190526
45169CB00009B/3295
* 9 7 8 1 5 3 7 1 0 0 7 6 0 *